How's & Why's

of

AN UNEXPECTED UNIVERSE

Sehdev Kumar

L.G. PUBLISHERS DISTRIBUTORS

First Published, 2013

ISBN 978-81-910382-5-5

Published by
L.G. PUBLISHERS DISTRIBUTORS
49, Gali No. 14, Pratap Nagar
Mayur Vihar Phase I, Delhi 110 091 India

Printed at
Mudrak, 30 A, Patparganj, Delhi 110 091

In Memory of

My Unlettered Mother

If seeds in the black earth
Can turn into such beautiful roses,
What might not the heart of man become
In its long journey toward the stars?

– G.K. Chesterton

Contents

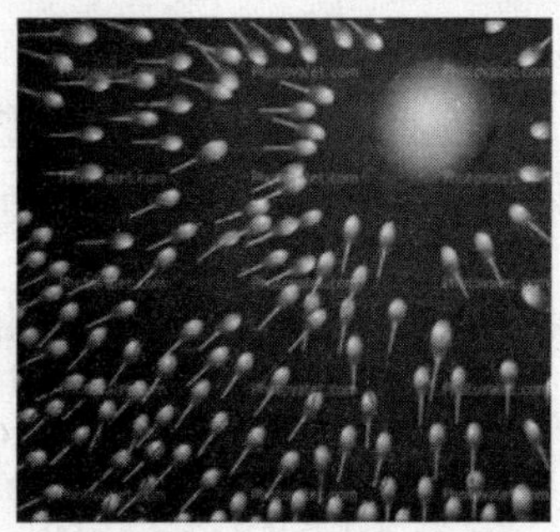

Preface

The How's and The Why's have been the warp and weft of the human exploration from the beginning of time. In our fairy tales and myths, in our poetry and epics, in our laboratories and temples, in geometrical theorems and algebraic formulas, in mathematical equations and chemical formulations, for millennia we have wondered about many why's and many how's.

How does a Monarch butterfly find its way from Point Pelee in Ontario all the way to a remote hill in Mexico? Why do Emperor Penguins emerge from the icy ocean and tread a few hundred miles to breed in a very specific spot that is isolated, icy and most inhospitable? How and why does a child born of kind and gracious parents become a vicious assassin? Why do bad things happen to good people? How do words of a poem written hundreds of years earlier reverberate in the heart of a forlorn lover? Why does a limb, so long lost, seem so real?

These, and a thousand such questions, and the answers they have elicited, continue to enrich our lives in myriad ways. Yet new questions arise eternally. In George Bernard Shaw's play *Back to Methuselah* (1921), the Serpent says to Eve in the Garden of Eden: "You see things; and you say, 'Why?' But I dream things that never were; and I say, 'Why not?'" The play spans ages from Adam and Eve to 31,000 AD.

In the last thousand years, even in the last fifty years, the human story has changed so fast and in such unexpected ways that any prediction about the future seems somewhat spurious. Yet, over millennia, one human trait seems not to have changed very much: We still ask 'Why?', and we still sometimes counter 'Why not?'

Yet for all that daring probity, we are often so moribund in our perceptions that anything out of the ordinary seems to threaten us. Four hundred years ago, in 1609, in Venice, Galileo Galilei was so enthralled by the sight of the 'beautiful and delightful body' of the moon as he saw it through his telescope that he wanted one and all to witness it. Many came to see the new wonders through the new instrument, but one person he so wanted to come refused; Galileo complained bitterly against this "principal professor of philosophy who I have repeatedly and urgently requested to look at the moon through my glass [telescope], which he pertinaciously refuses to do."

Often we are so encaged in our perceptions that to look at something different, or to look at what is familiar differently, becomes a challenge and a threat. Yet without new perceptions, how little do we evolve.

One man hangs a sign in his bookstore: "Be not inhospitable to strangers lest they be angels in disguise," and then, over the decades, opens his doors to over 50,000 men and women from all over the world who have come to explore the city of light.

Five hundred years earlier a man of genius regularly buys caged birds at a market and sets them free.

"How could He do this to us?" cries a man whose five children lie buried under the rubble of a home in a devastating earthquake. "There is no God," he moans. A woman passing by a pyre of burning bodies pulls out a copy of the Bible and throws it into the fire. "Why me?" – "Why me?" – an unbidden, excruciating cry of the spirit, wrenched out of agony, forged not in pity but in outrage.

Another man, in his search from Quantum to Cosmos, declares: "My goal is simple. It is complete understanding of the universe, why it is as it is and why it exists at all."

These are some of the nodal points for meditations in this book; there are far more questions here than answers. That is how our journey must go on.

Part I

Wandering in An Unexpected Universe

If you do not expect it,
You will not find the unexpected,
For it is hard to find and difficult.

– Heraclitus, 500 BCE

1

Angels in a Bookstore

On the Left Bank in Paris, in the long shadows of the Notre Dame, there is a quaint bookstore called 'The Shakespeare and Company,' once described by its owner and keeper, George Whitman, as "a socialist utopia masquerading as a bookstore." Indeed, it is a bookstore and a library, and a shelter for roving and impoverished young writers who still gravitate to Paris for inspiration and ideas. Through its many incarnations at many locations, going back to 1920s, the shop is associated with many legendary figures of literature, such as Ernest Hemingway, Ezra Pound, Scott Fitzgerald, Henry Miller and James Joyce, who nicknamed it 'Stratford-on-Odéon'. Poor, unknown young writers still come here from far and wide for some solace and for shelter for a night or two, or more, and then move on, cherishing its conviviality and its chaotic ambience.

One autumn afternoon in 1998, I was here at Sunday Tea, being warmly received by the gentle octogenarian George Whitman – a distant cousin, some said, of Walt Whitman.

I was a writer, but I was hardly starving. I had no special claims to anything; I had not come here to seek shelter, or to find a specific book. But I was drawn to this place knowing that this 'socialist utopia' had provided, over many decades, beds and pillows – such as they were – to over 50,000 men and women from all over the world; all it had ever asked of them was that they make their bed in the morning, and "help out in the shop, and read a book a day."

Painted above a doorway that leads to a rickety stairway to the second floor, and supported precariously by a stack of books that reaches gloriously from the floor to the lintel, are words that render a special illumination to this city of light: "*Be not inhospitable to strangers lest they be angels in disguise*."

Who knows what angels are, or if they are anything at all other than some supposed spiritual apparitions with some unexpected messages from a supernatural universe? For millennia, people in many lands have spoken of them as bearers of good tidings, particularly in moments of some unbearable anguish. They come, they say, when one least expects them; they come as a faint light, as a whisper, as a whiff. They come in all seasons – in spring, in autumn, in winter. They always come as strangers.

Strange unexpected encounters, thoughts and events are part of life of all of us. Sometimes they are more than a matter of chance; they seem providential; they give a new and unseen bend to our river of life; we don't know – in fact, no one knows – how and why they happen.

It seems to me that even as we get to know many things, beyond the roving and penetrating eye of science the universe is still filled with strangers, with strange things, strange thoughts, strange forms, and strange premonitions. How to be hospitable to such a strange universe as though it were indeed brimming with angels with infinite messages? "My own suspicion is," said the great British geneticist J.B.S. Haldane, "that the universe is not only queerer than we suppose, but queerer than we *can* suppose."

Today, on a cold winter night in the Doon Valley in the Western Himalaya, I am seized with the thought of angels and with the sign that I saw in the Shakespeare and Company bookstore in Paris more than a decade earlier. For, in truth, less than a year ago, even in my wildest dreams, I had no idea that I would be living in this valley; it all happened through a chance encounter, which no laws of probability can explain away.

"If you do not expect it," said the Greek philosopher Heraclitus in sixth century BC, "you will not find the unexpected, for it is hard to find and difficult." I don't know what could have led me to expect it.

Here in India, in a population of 1200 million people, there are some 800 million mobile phones, possibly more. Not long ago, not many people in the country had any phones; even those who had them, could hardly use them very reliably. But now everyone is talking and talking and talking. Not only that, these smart phones receive and transmit messages, pictures, and images, hundreds of millions of them every second. And unlike the traffic jams that are here everywhere, these messages don't collide into each other at all.

I feel bewildered that these spaces all around us, which seem so manifestly empty, are really not empty at all; they are in fact filled with a million, and a billion thoughts, words, images and messages, all criss-crossing at dizzying speeds. Where in this seemingly empty space are all these utterances and images hiding, and how they are formed and transformed and re-formed before they are articulated, I wonder.

In the great void that prevails in the vast inter-stellar spaces, even in the cavities of our own hearts, and in each atom itself, are there some messages that we do not hear, or perhaps cannot hear?

Somewhat grudgingly I am beginning to believe that there may be angels after all; maybe they are indeed sending us some encoded messages from somewhere, enfolded in the great void in the universe. Or in many universes, as is being surmised.

Like many others I have known moments and days of anguish, though I have never come close to death. In deep anguish, or in the throes of death – in a war, in an earthquake, in a raging fire, in an explosion or an accident, or in the face of a terminal disease – do we not sometimes imagine what lies on the other side of the abyss? Are we not then driven to seek some solace from some transcendental force?

It is one of the great tragedies of life that only a few among us have stared deep down that abyss of despair and have come out intact, and had the gift to speak about the light that guided us in that dark hour. One of the most exquisite examples of such a rare person was Viktor Frankl. He was an inmate in a Nazi concentration camp where he experienced, like all other inmates, horror and atrocities in a manner, and on a scale, unknown in human history. In 1946, based on his experiences and on his reflections, he wrote *Man's Search for Meaning*, a book that has become one of the most influential books in the world. His words are worth repeating a thousand times, for such truth is revealed in no science, in no ideology and in no philosophy; it can come only from an angel:

> A thought transfixed me: for the first time in my life I saw the truth as it is set into song by so many poets, proclaimed as the final wisdom by so many thinkers. The truth – that love is the ultimate and the highest goal to which man can aspire. Then I grasped the meaning of the greatest secret that human poetry and human thought and belief have to impart: *The salvation of man is through love and in love.* I understood how a man who has nothing left in this world still may know bliss, be it only for a brief moment, in the contemplation of his beloved. In a position of utter desolation, when man cannot express himself in positive action, when his only achievement may consist in enduring his sufferings in the right way – an honorable way – in such a position man can, through loving contemplation of the image he carries of his beloved, achieve fulfillment. For the first time in my life I was able to understand the meaning of the words, "The angels are lost in perpetual contemplation of an infinite glory...."

For millennia we have tried to probe the mystery of our existence in our laboratories, in our books of learning, and in our temples. We have understood a great deal but each one of us still awaits an angel who will whisper to us some words of solace when all else is silent and dark. An angel who would tell us of the transcendent reason behind the apparent cause.

Not long ago, at a festival of Quantum-to-Cosmos at Perimeter Institute in Waterloo, a very celebrated scientist, Stephen Hawking, declared: "My goal is simple. It is a complete understanding of the universe. Why it is as it is and why it exists at all."

As an eloquent voice of science, perhaps Hawking could unravel the mystery of the universe to us. But somehow I doubt it, unless, of course, the universe he wants to understand completely is dead, unchanging, uncreative, and without consciousness. The wonder of the universe is that it is forever making itself anew, if not among the stars and the galaxies, then certainly here on earth where we experience it with our every breath. What we discover is that, despite its numerous laws and cycles, it is an unexpected universe, full of surprises and astonishing turns of events and thoughts.

This unknowability is the puzzle at the heart of our existence. This is when we invoke angels; we imagine them and create them. Then they come to us in many forms: sometimes as a whisper, sometimes as an embrace, sometimes as a song. Sometimes as a dusty book in a dishevelled bookstore. That's how angels must have come to those young budding writers from far and wide as shelter and solace in the Shakespeare and Company bookstore in Paris over many decades.

There is no law of nature that can explain such things. These angels in disguise always seem to hover at the edge of nature, in some penumbral region that appears so obviously empty, sending messages to some forlorn ear that may be momentarily tuned to receive them. To such an ear these angels are as real as the SMS message it just received on its iPhone.

I wonder, over the decades what messages the great geniuses of literature received in that dusty bookstore in Paris to create such firmament of imagination.

For thousands of hungry writers who found shelter at the Shakespeare and Company bookstore, the greatest angel of all must have been George Whitman himself.

In 2011, at the age of 98, on a cold night in December, George Whitman moved on to some other cloud.

❖

2

Six Embossed Points: The Alphabet of the Universe

Two hundred years ago, in 1809, was born a man in France whose invention brought a revolution to millions of people all over the world. His name was Louis Braille; he is the only man in history who could be called the father of a new script for more than two hundred languages; the script for reading and writing a language – including the language of music and mathematics – that he invented is named after him, Braille; his script and method are used by all those who are deprived of eyesight, the most precious of nature's gifts.

At the age of three, when toying with a leathering tool, Braille was blinded in one eye by an awl. And then through infection, the second eye was lost too. By the age of five he was totally blind. But goaded by an army captain who was creating a secret script for use in war, Louis set forth to transform the sounds of language and the letters of the Roman alphabet into something tactile. Thus a new metamorphosis took place: from sound, not to sight but to touch. A touch now conveyed nuances of meaning and thought; a new script and language for the blind had come into being. Through one's finger tips, the vast universe of thoughts and emotions was now suddenly made accessible. Though there is no one person who can be said to have given us the first writing script some 5000 years ago, or any number of other scripts that have emerged over millennia, Braille can unmistakably be commemorated, as he was in many countries in postal stamps and coins in the year of his second birth centenary in 2009.

Six dots. Six embossed points. Six bumps in different patterns, like constellations in the sky, and a whole new universe of thoughts, ideas, poetry, mathematics, and music was laid bare. In early nineteenth century, when the world of the blind was doomed to be dark forever, at the age of fifteen, a sickly poor boy gave a very special gift of light to the world.

The wiggles and squiggles of black ink on this paper, which capture our most delicate feelings and profound thoughts, and which many of us now take for guaranteed, began to emerge only about five thousand years ago. Before that, for eons, we had a world of many sounds, and of many spoken languages. But to write that world down and to read it, to transform a thought and a feeling into sound, and then into a script, is a grand revolutionary step forward, far grander in our story of civilization than possibly anything else. Yet in early 19th century Europe, in the Age of Enlightenment, only a few could read and write. In the France of 1809, when Braille was born, more than half the population could not sign its name.

Blindness, it must be said, is both a physical affliction and a state of mind, and of heart and soul. How often we have eyes to see yet don't see. We are blinded by our arrogance or fear, by sloth or insecurity. "Eyes and ears are bad witnesses for men," says Greek philosopher Heraclitus, "if they have barbarian souls." With all the literacy in the world, we can still continue to be barbarians.

I often wonder how does it feel to be blind in a garden bursting with colours? To be deaf in a universe resounding with a thousand symphonies?

To be unlettered in a home full of books? To be hungry, and be cordoned off from a banquet table?

My mother was utterly unlettered, and remained so for all eighty years of her life. "For me *kala akshar bhains brabar*," she would say with great pain: "the black letters in the book are no different than the skin of a buffalo."

She knew the buffalo well: in her youth as a young bride, she tended to them for milk and fuel, and they were part of the wealth of the family. But in the black hard skin of the buffalo, what could one read and be illuminated by it?

I marvel at Louis Braille's creation: how through the touch of six embossed points, a world is revealed and made manifest. Yet I wonder if there are other points in the universe one has not learned to touch and be illuminated by. This thought came to me when I saw Iranian filmmaker Majid Majidi's film *The Color of Paradise* about an 8-year old blind boy Mohammad. For his widowed, impoverished father Mohammad is a burden, a good-for-nothing son, forever to be taken care of. The father's anxieties and fears have blinded him to his son's special gifts to read and write Farsi in Braille, as well as a thousand other embossed points in all of nature – in the petals of flowers and the colours of plants, in the human faces, in the scales of fish, in the feathers of birds. His father, much like most of us – though fully 'normal' and endowed with sight – only hears wild frightening sounds in the forest and on the sea shore. Mohammad, on the other hand, hears the music of heavens in the sound of birds and waves.

With his hands the blind boy can *see* the colours of paradise.

When we are encaged in our own bloated anxieties, we touch nothing and the whole thunderous Niagara of life slips away through our fingers. When we blame our suffering on others, then what is beautiful and is ours, flees away from us, making us ever more impoverished.

Much before we humans invented alphabets for our various languages, and long before we became writers of our thoughts and feelings, we were, and forever continue to be, readers of what the great poet Coleridge called 'the mighty alphabet of the universe'. The signs of being and becoming, of transformation and discovery, of the heaving earth and the bursting seeds, of the path lost in the woods, 'of the way up in the torn clouds', are evident in all places, in the falling of the leaves no less than in the migration of Monarch butterflies thousands of miles away.

There is a great firmament on earth and in heaven, and in every human heart. This is what the blind boy Mohammad could touch and read. And this is where so often we remain utterly unlettered.

"The Lord, whose oracle is at Delphi," said Heraclitus, "neither seeks nor conceals, but gives signs." Each one of us deciphers these signs, and renders meaning to them, in accordance with what we can *read*. Without the gift of this literacy, we are all somewhat blind. The luxuriating world of sights and sounds then seems no different from a vast stretch of a buffalo's dark skin, without any embossed points for our touch.

A world bereft of meaning!

My mother was unlettered, of course. But I often wonder in what ways she had learned to touch the universe. And in what woeful ways, I and my learned friends remain unlettered.

❖

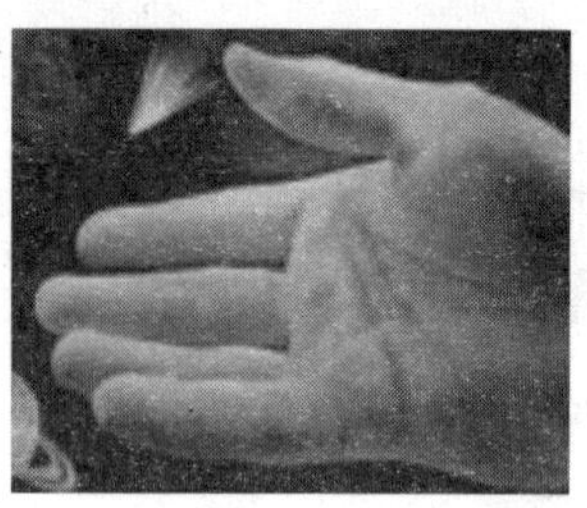

3

The Invisible Hand of God

In 1755, on November 1, on All Saints' Day, Lisbon was hit by a great earthquake, followed by a devastating tsunami, bringing destruction and chaos on a scale not known to the people of Portugal before. That such wrath of God would be expressed on a most sacred day shook the faith of many: "Why would God punish his people so?" they wondered. As great philosophers and intellectuals – Voltaire, Kant, Rousseau and others – debated about the meaning and purpose, and the causes of such an earthquake, the immense natural catastrophe proved to be the watershed for the birth of modern seismology.

Christian theologians, nevertheless, struggled to justify the natural tragedy as an irrefutable act of God, and as an *intention* of God, and to explain it in terms of some guilt Lisbon's inhabitants had cumulatively incurred to merit such dramatic retribution. A few months earlier, on the orders of the King of Portugal, the Portuguese armies had massacred thousands of natives and missionaries in Paraguay and in other countries in South America. Thus there were prophesies for more disasters in the next few days and weeks, which produced more terror in the populace, even as the learned derided such prophesies. But the learned too were left with much to explain: the eighteenth century had been a period of philosophical optimism. The discoveries of Newton and Leibniz had promised rational explanations for the order of a nature that was in harmony with the designs of God. What possible harmony or design, many were forced to ask, lie in the devastating earthquake? Alas, there were no answers to be found; one could console only by saying: "Men die, but men *always* die. God remains good."

Some two hundred years later, on January 15, 1934, in India a severe earthquake rocked Bihar, unleashing extensive devastation over a vast area; 650 kms away, in the city of Kolkata, the tower of St Paul Cathedral was destroyed. The earthquake killed more than 30,000 people at a time when India was engaged in a vigorous movement for independence under Mahatma Gandhi. The natural causes for an earthquake, and the ability to predict one, were little known then – and are still barely known. However, grieved as he was at the fate of hundreds of millions of outcastes and untouchables in India, for Gandhi the devastation was a sign of wrath of God. "A man like me," he argued, "cannot but believe that this earthquake is a divine chastisement sent by God for our sins" — in particular the sins of untouchability. "For me there is a vital connection between the Bihar calamity and the untouchability campaign."

In India in 1934, there were few voices as eloquent and as lucid as those of Jawaharlal Nehru and Rabindranath Tagore about what understanding nature through the eyes of science meant. Nehru was shocked by Gandhi's assertion about the earthquake: "If the earthquake was a divine punishment for sin, how are we to discover for which sin we are being punished?" he wondered. "Why did not the earthquake visit the land of untouchability itself? Was it a judgment on the prevailing zamindari system since many rich land owners had suffered losses in the earthquake?" he asked rhetorically. "Could the British rulers interpret it as a divine punishment because Bihar had been taking a leading role in the freedom movement?"

Tagore, who abhorred untouchability equally and had joined Gandhi in the movement against it, protested against this interpretation of an event that had caused suffering and death to so many innocent people, including children and babies. He also hated the epistemology implicit in seeing an earthquake as caused by an ethical failure. "It is," he wrote, "all the more unfortunate because this kind of unscientific view of [natural] phenomena is too readily accepted by a large section of our countrymen."

Tagore argued with Gandhi that physical calamities must have their origin in physical facts. He believed in the inexorability of the natural law in the working of which God himself never interferes. He felt that our own sins and errors, however enormous, could never have enough force to drag the structure of creation to ruins.

In 2010, another great tragedy was unleashed by an earthquake, this time in Haiti. Invariably there were millions in Haiti and elsewhere who saw in it God's wrath against human depravity. It is not a mere act of nature, they said, but retribution of God against those who are morally decadent. American Evangelist Pat Robertson called this as a curse on a country that had dared declare the first black republic in 1804 and overthrow the French rulers, who had through slave trade, like the Americans and the British, peopled Haiti with black slaves from Africa.

"Who is corrupt and who is being punished?" one wondered. The Justice Ministry and the Cathedral were destroyed. The National Palace and the United Nations headquarters were destroyed. "Redeem yourselves! The end of the world is near!" shouted a man. "Haitians need to reinvent themselves and find a new path to God," implored Rev. Eric Toussaint outside the ruined cathedral.

"How could He do this to us?" cried a man whose five children lay buried under the rubble of a home. "There is no God."

A woman passing by a pyre of burning bodies pulled out a copy of the Bible and threw it into the fire.

How to make sense of such a horrific tragedy? Who is to be blamed for it? If God is good, why does this destruction come? If it is an act of nature, how is that some are saved, and some are killed? Does nature have no intention? Is there any moral principle imbedded in nature that resonates with human sense of good?

There is a long and lingering tradition in human history for seeing events in nature reflecting our own being and becoming on this earth; we find it difficult to believe that human mind may not be entwined with the intentions of nature. What is unexpected, unknown, and mysterious – and there is much that is so, and would probably always be so – conjures up in us visions of the Great Mystery. Over millennia, nature has revealed many of her secrets to us; we have inserted our probing hand in the hidden and dark recesses of the unknown. We have touched much and have come to know a good deal. Yet when the great earth shakes we don't know who will die and who will live.

Is this uncertainty inherent in nature? Or is it unknowable to us humans? For, with all our learning, we still don't know how to respond to the despair of the sorrowful man whose five children are buried under the rubble of a home, and who is asking: "How could He do this to us?... There is no God."

4

How Natural is the Sacred?

In May 1945, as the atomic bombs were becoming a possibility as part of the Manhattan Project, very significant strategic questions were being raised in Washington: which cities in Japan should be the target of these new weapons? Tokyo itself was ruled out because it had already been firebombed for months. In order to see and study the full impact of these new weapons, the cities for atomic bombardment had to be relatively untouched by war. Hiroshima and Nagasaki were finally chosen. Yet one other city that was considered was Kyoto.

Kyoto was the ancient capital of Japan; it has some of the most magnificent Buddhist temples, some dating back a thousand years. It so happened that Henry L. Stimson, Secretary of War in the Roosevelt and Truman administrations, had spent some time in pre-war Japan, and he knew what these temples meant to the people of Japan. To bombard Kyoto, he argued, would be creating an enemy who would haunt America for ever; Kyoto was the very soul of Japan; it was sacred. And thus the city was spared.

A year earlier, as France was being liberated, to smoke out the hiding Nazi soldiers from inside the Chartres Cathedral, liberating American armies talked about bombarding it. The cathedral was, however, saved because someone had whispered what the great cathedral meant to the people of

France despite their pronounced anti-religious stance. Chartres was the symbol of French heritage; it was sacred. Any damage to it, whatever its tactical purpose, would invite the wrath of a proud people even at the moment of their liberation.

Forty years later, in 1984, when Prime Minister Indira Gandhi sent soldiers into the Golden Temple in Amritsar to flush out the militants, the Sikhs were outraged; their most sacred place had been assaulted. Less than six months later the Prime Minister was gunned down by her own bodyguards, and Canada became the target for the worst terrorist act in its history in the downing of an Air India plane in 1985.

How does a building or a place get so imbued with significance that it becomes a symbol of utmost sanctity to a people? How do bricks and stones assume such grandeur that they become worth dying and killing for?

One wonders if these symbols must always be religious to be 'sacred'. When a flag of a nation is denigrated or burnt, it can cause festering anger in a people, because, as a symbol of a nation, it has sanctity. A few years ago, when a group of people in Belleville, Ontario trampled a Quebec flag, the insult fuelled a rallying cry for separatism. When books are burnt, or a library is destroyed, it is seen as an assault by many who consider a book a sacred symbol of freedom of expression. A vase that contains the ashes of a deceased child is sacred to a mother; as a depository of memories, on a winter night, in the hush of the falling snow, she can sculpt from these ashes a mirthful world that had ended so unreasonably and unexpectedly in a moment.

In 1170, when Archbishop Thomas Becket was murdered in the cathedral at Canterbury by King Henry's soldiers, it was a betrayal of friendship and of a word given. It was an act of sacrilege, as all betrayals are; for penance, the King had himself flogged.

For many families and communities, for millennia, the honour of women has been sacred; if they colluded with the enemy, as some French women, for instance, did with the Nazis, their hair was shorn off and they were paraded on donkeys, for having brought shame to France by sleeping with the enemy, as depicted so poignantly in Alain Resnais's film, *Hiroshima Mon Amour*. In the film, *Zorba the Greek*, a widow is stoned to death by the villagers, in the presence of the priests, for consorting with an English man in preference to a local suitor.

Human life, all through its long and tortuous evolution, has always been imbued with something that is 'sacred'; sometimes it is a place, or a monument; sometimes it a memory. Sometimes it is an idea; sometimes it is a belief, or a cherished value. Sometimes it is a word given, or a promise made. Sometimes it is a friendship. Sometimes it is love. Sometimes it is trust and loyalty. Sometimes it is a work of art.

On April 26, 1937, at the behest of the Spanish Nationalist Forces of General Franco, the village of Guernica in Basque Country was bombarded by German and Italian warplanes. The horrors of the tragedy seethed so deeply in the Basque memory that every young man pledged it to be his sacred duty to seek revenge. Pablo Picasso captured this ferocity in his painting *Guernica*; it became a sacred symbol of assault on the innocent everywhere, in every age, and under every brutalizing force of repression. Picasso's painting travelled the world but it never went to Spain untill after the death of Franco in 1975. Now it resides in Museo Reina Sofia in Madrid, in a bulletproof case, still a symbol of the vulnerability of innocence in a desecrating world.

There is no rationale or science for the sacred; the sacred is bred in the human bones, possibly in the bones of all creatures in the animal kingdom. In commemoration of what can only be called sacred, Neanderthal man, 100,000 years ago, is known to have placed stones on burial spots in awe and memory. When elephants weep for the loss of their calfs, or emperor penguins endure extraordinary suffering to bring new life to fruition, I wonder, if they too express a sense of the sacred.

We have made rivers and mountains sacred; we have worshipped trees and rocks; for millennia, we have offered flowers and rubies for adoration and in gratitude to gods heard about or imagined; we have travelled the world, across deserts and high passes, to satiate a deep-rooted impulse for the transcendent.

We are symbol-making creatures; we imbue some symbols with significance and sanctity, and make them integral to what we believe is the centre of our being, and what renders meaning to our lives. A flag is a mere piece of cloth dyed in specific colours; to a stranger it has no meaning. A book is a mere bundle of papers with ink on them, but when it is burnt because of what it dares to say, it touches our liberal conscience; it violates our sense of the sanctity of freedom.

In our multicultural world what is sacred to one person or one community may be a mere symbol of obscurantism to another, and hence an easy target of satire and derision.

In 1776, when the United States was forging a new nation, the like of which truly had not existed anywhere in human history, one of its bold and revolutionary founders, Thomas Jefferson, penned words for the American Declaration of Independence, which to this day are as stirring and bold as they were when first written more than 200 years ago:

> We hold these truths to be self-evident, that all men are created equal, that they are endowed by their Creator with certain unalienable Rights, that among these are Life, Liberty and the pursuit of Happiness.

This sentence has been called "one of the best-known sentences in the English language" and "the most potent and consequential words in American history." Yet, it must be said, that what it proclaims to be 'self-evident' is based on no sociological or scientific study, nor is it in accordance with any laws of nature, nor was there any agreement who the Creator was.

This is the fervour of revolutions, whether in one's own life or in that of nations; they require more than pedestrian reason or rationale to do what they aspire to achieve; in their grand sentiment, like all romantics, they believe that God is on their side and the whole universe sings with them.

A hundred years later, in 1886, the grand 151-foot Statue of Liberty, a gift from the people of France to America, was installed in New York Harbor. As millions of new people came from Europe to the shores of this country of new dreams, this is the first thing they saw of the new country: a grand woman, with a torch in her hand, promising liberty. However, what is often not known is that as the pedestal and the statue were being erected, in 1883, a 34-year old Jewish and socialist poet, Emma Lazarus, wrote a poem, 'The New Colossus', to welcome the statue. She called her 'The Mother of Exiles'. Twenty years later, in 1903, words from her poem were inscribed on the pedestal of the statue, which have become as grand and as prophetic as the words of the American Declaration of Independence:

> *Give me your tired, your poor,*
> *Your huddled masses yearning to breathe free,*
> *The wretched refuse of your teeming shore,*
> *Send these, the homeless, tempest-tossed to me,*
> *I lift my lamp beside the golden door!*

Again, one could ask, where do these words come from, and what dreams of a new nation are woven in these words, and how they have become a tapestry of welcome – a sacred promise – to people from many lands, and with many woeful tales.

Charles Darwin called sympathy 'the noblest part of our nature', which could not be checked 'even at the urging of hard reason'. It was not a 'scientific' observation, based on any facts. It was part of his dream of what he considered to be a good and noble life; such a dream invariably transcends both science and nature; its origin lies in the human experience and in the human yearning for a higher life. This is how the sacred is made natural.

Again, this is how on December 10, 1948, the Universal Declaration of Human Rights was adopted, promising an unknown new dignity for hundreds of millions of people all over the world for the first time, irrespective of who they were. Such a promise arose not from any discoveries in science or in a moral void. It emerged, in the age of science and after the most gruesome war in the human history, and from the ashes in concentration camps in the Holocaust. It emerged from hundreds of years of national, tribal and religious wars in every continent; it emerged from the genocides by imperialist European powers of indigenous people in the Americas; it emerged from the brutality and inhumanity of slavery and slave trades over centuries; it emerged from the pain and scars of serfs, indentured and bonded labour; and from a thousand other brutalities that have been heaped on the marginalized, the poor and the defenceless for millennia. Above all, it emerged from unexpected turn of historical events, and from the turmoil of human conscience, from the sympathy and concern of men and women of imagination and goodwill, and from our inherent desire, so often neglected and smothered, to live in a moral world.

The sacred permeates all of human life in a myriad ways, and renders meaning and dignity to our individual and communal selves: a haunting memory of a friend is so personal, so very much our own; a trust is merely a sentiment, so easy to betray. A value or a belief is no more than a guiding star on a dark night, so easily forgotten, so conveniently ignored. But in this exquisite subjectivity of the sacred for an individual, or a group, or a nation, or a people, what is most essential to our being, shines through; it is then that we see our true human-ness; it is then that we have a glimpse – however faint, and however ephemeral – of the transcendent, that Self that encompasses our self, and yet alludes to something far greater.

Just as the words 'Humanity' and 'Crimes against Humanity' mean far more than any one human being, the Sacred too refers to all that we truly, unquestionably, and most ardently, hold dear. It gives sustenance to our soul and to our being.

The Sacred is beyond reason, and yet it becomes the reason for our being.

❖

5

Making of a Masterpiece

"The world," argues Steven Weinberg, Noble Laureate in physics, "can always use heroes, but could do with fewer prophets." Prophets of religions, he says, are deemed infallible by their followers; their words and authority are never to be questioned, and their edicts are to be obeyed with an unflinching blind faith. The heroes of science, on the other hand, are admired for their insights and ideas, all of which are always open to challenge. With the spread of science and scientific ethos – "independence of mind and openness to contradiction" – prophets and faith in religion, he believes, are falling by the wayside. "One of the great achievements of science has been," he argues, "if not to make it impossible for intelligent people to be religious, then at least to make it possible for them not to be religious. We should not retreat from this accomplishment."

Dr. Weinberg has spoken passionately and extensively against obscurantism, supernaturalism and promise of afterlife and other kinds of religious mumbo-jumbo, and he has rejected any notion that religions are the basis of any social or personal morality. "Religion is an insult to human dignity," he asserts. "With or without it you would have good people doing good things and evil people doing evil things. But for good people to do evil things, that takes religion."

It is impossible, indeed utterly unconscionable, to deny the evil that has been done for centuries in the name of religion, by followers of one sect against another, by the dogmatists and the fundamentalists, by the pig-headed and the literalists, by the gurus, priests and the mullahs, by the purists and the self-righteous, and by the mighty and the misguided. Religious leaders have endorsed slavery and untouchability, murders and wars, burning of widows and of 'witches', sacrifice of children and women as part of God's order. But ... and it is a very big But ... there have also been unceasing voices in every age and in every religious culture of love and compassion, of tolerance and mercy, of unity and oneness of life. "*Ek noor se sab jag upjaya* ... From one light all has come to be," sang Kabir five hundred years ago – before the advent of Scientific Revolution. Sa'd al-din Mahmud Shabistari declared in the 14th century Persia:

> *What are "I" and "You"?*
> *Just lattices*
> *In the niches of a lamp*
> *Through which the One Light radiates.*

These are the voices that have pleaded again and again, and for millennia, for turning swords into ploughshares, for offering the other cheek to an offending hand, for feeding the hungry and consoling the lonely, for piercing through the delusions of *maya* to see the face of unadorned reality, for daring to experience the oneness of all things, and for love to manifest itself as the most elemental force in the universe.

It is these voices – however rare, however faint – that seem to keep the religious impulse alive, with all its convoluted pathways, vulnerabilities and seductions.

Just a few weeks before the passing away of the Indian sage Sathya Sai Baba on April 24, I was at his ashram in Puttaparthi in Andhra Pradesh. This was my first visit; I had gone there not as a follower but as an observer of a very remarkable phenomenon. Here was a man whom many in India addressed as 'Bhagwan' or 'God-like'; the media referred to him and other such spiritual teachers as 'God men'. Amongst thousands of men and women who poured in at his ashram every day, there were hundreds who came from far off lands, and who were Muslims, Jews or Christians, and who were drawn, they often said, because the place was 'filled with love'. The Baba presented no erudite philosophy, no intricate rituals to observe, no elaborate beliefs to uphold. He spoke little, wrote even less, preached hardly at all. Once he was known to perform certain miracles, but with years that was all gone. All he asked those who came for his *darshna* was to live a life of love and selflessness. It was a message as old the mountains, but rarely heard, rarely practiced. Yet it was still eternally beckoning, forever resonating in the hearts and minds of those who were perhaps gullible, or still hungry for such a message. Like a romantic tale – of Romeo and Juliet, of Laila and Mujnu – it is a message that is very old and worn out, but it does not fade away. Like some great work of art. Like a poem. Like a piece of music. The Baba, it seemed to me, was part of a long tradition of India, perhaps of the world at large, that touches upon the eternal that is not new, but that renews itself eternally.

"No creature in the world," observes evolutionist Loren Eiseley, "demands more love than man; no creature is less adapted to survive without it."

Many years ago, Michelangelo's great masterpiece, *The Pietà* was brought to New York to be exhibited in a world fair. Carved in 1498-99, this was the first time that this exquisite work of Renaissance art was making a journey abroad. The fame of the masterpiece, already immense, had travelled far and wide; it attracted literally millions of visitors to the fair, one could view the marble sculpture for only a few minutes from a moving platform. The 'ohs' and 'ahs', and the flashing cameras were all part of a quick and hurried *darshna*. We felt lucky to have witnessed something so grand and so eternal.

Is it possible, I wonder, that a man becomes Bhagwan, when he so 'carves' himself that he is transformed into a masterpiece, a 'piece of the Master'? Is this not the true and real evolution of each one of us that is light years away from all Darwinian evolution? It is not so important what kind of monkey we came from, someone once remarked, but what kind of monkey we are to become.

It is this self-directed evolution – of slow and arduous chiselling – that has been at the heart of spiritual quest, despite all the horrors of religions. This journeying between 'being' and 'becoming' and 'being' again, may be the 'religious impulse' that may take man beyond, what is often so presumptively called 'natural'; it may possibly take him into the realm of the sacred.

For Michelangelo:

The best of artists hath no thought to show
Which the rough stone in its superfluous shell
Doth not include: to break the marble spell
Is all the hand that serves the brain can do.

Sri Ramana Maharshi was a much loved Indian sage in the last century who spoke to others mostly through silence; the Silent One just sat there, like Michelangelo's *Pietà*, as thousands thronged for his *darshna* from far and wide at the foothills of Arunchala. For his followers, he was a Master, a human masterpiece, a piece of the Master. Through self-reflection, he had seen the face of reality; he had discovered the nature of the self. He used no microscope or telescope; his gaze had been directed within. He said to those who came to him, in few words and in deep silence, that self-discovery was the real road to freedom. Freedom from what? one may ask.

Freedom from craving? Freedom from lovelessness? Freedom from ignorance? From fear of the void?

"In our struggles and craving for freedom, what is the meaning and value of such freedom?" one must ask.

What did *The Pietà* say to those who came to see it by the millions at the World Fair in New York? And what does it say to those heaving crowds who come to see it in hordes in St. Peter's Basilica at the Vatican every day? That human hands and dazzling imagination can create a work of an extraordinary beauty and power; that the silent message of pity and compassion can travel across continents and across centuries, and yet still be barely heard.

"What is *The Thinker* thinking?" one might wonder, and find no answer.

A people can be mesmerized by the grandeur of music and of art, and still be barbarians and destroyers of life. The dragons that lie coiled in the human heart can be spurred by ideologues as furiously as by religious zealots, and no less by the relentless and fierce powers of science and technology. They may also not be lulled by any music or by any art. "Our scientific power has outrun our spiritual power," Martin Luther King once exhorted. "We have guided missiles and misguided men."

As a distinguished physicist, Steven Weinberg is "professionally concerned with finding out what is true, not what makes us happy or good." But the latter – "what makes us happy or good" – is surely not a trivial pursuit. It may require looking around, and within; it may even persuade one to question 'what is true', and why is this truth so worthwhile.

For some it may even require asking: "What makes an ordinary stone into a masterpiece?"

❖

6

Truth: 24 Frames per Second

In the heyday of the New Wave of French Cinema, radical filmmaker, Jean-Luc Godard, once wrote that "A film is truth 24 frames per second." In his own provocative manner, Godard was referring to the power of films, and how each one of the 24 frames projected on the screen every second, was imbued with truth. Undoubtedly, over the decades, some films have had the extraordinary power to stir us – or amuse us, provoke us, educate and inform us, and bring many tears to our eyes and much laughter. Nevertheless, a film is an optical illusion that is created because of our perceptual limitations.

"An optical illusion," the great German poet Goethe insisted, "is an optical truth." And so it is! Truth, however, is an all-encompassing word; Pablo Picasso declared that "Art is the lie that tells us the truth."

We have been engaged for a long time to find truth; some even 'Ultimate Truth'. Gandhi called his autobiography: *My Experiments with Truth.* And talked, like many others, that God is Truth, and Truth God. Isaac Newton, the greatest of scientists, though never a very humble man, wrote towards the end of his extraordinarily creative life: "I do not know what I may appear to the world, but to myself I seem to have been only a boy playing on the sea-shore, and diverting myself in now and then finding a smoother pebble or a prettier shell than ordinary, whilst the great ocean of truth lay all undiscovered before me."

In the most celebrated words in the Brihadaranyaka Upanishad 1.3.28,

Asato mā sad gamaya
tamaso mā jyotir gamaya
mṛityo mā amṛtaṁ gamaya
oṁ śānti śānti śānti

Lead me from untruth to truth,
From darkness to light and
From mortality to immortality.
Let there be eternal peace!

With our age-old search for truth – in science, in arts, in philosophy and in history – we have reason to ask anew: What is Truth? What are we asking to be led into? Those who were led into it, what did they find once they got there? And did truth set them free? From what? One yearns to hear real encounters, warts and all, without metaphysical pretensions.

Over the years, as I have struggled with these questions, I have come to believe that truth is shrouded in illusions and may be no more than an illusion itself, though it may appear in the light of the day, as solid as a rock, as transparent as glass, and as real as one's own hand.

An illusion, it must be said, is not merely an illusion; it only alludes to the limitations of our perceptions. It is a strange paradox: the world is and it is not; I think, therefore I am; I think, therefore, I am not. In each assertion there is an inherent contradiction; the point and the counter-point stand together. "What compels us to assume," asked Friedrich Nietzsche so disarmingly, "that there exists any essential antithesis between 'true' and 'false'?"

Death, for instance, is a certainty. Yet it is also an illusion. We speak of "Ashes to Ashes...Dust to Dust." To console the grieving, we speak of the immortality of the soul; we conjure up visions of the hereafter, of a celestial journey to unknown lands, of heaven and hell and of the last judgement. Yet as the funeral pyre burns, or as the casket is lowered into the innards of the earth, what is real for the living is the pain of going away;

the cessation of life – and the suffering it inflicts on the living – is real; all else is illusory. The smoke-filled air contains the dismembered molecules of the whiff of life that once experienced anguish or laughter, but now only renders a warm thrust to a kite overhead. The incantations of the priest about the imperishability of the soul are a lie – or so I believe it must seem to the wailing widow and the orphaned child. The soul may be immortal but a thousands bruises of the flesh, of my own and of others, make me human, and make me *real.*

Layer upon layer of our perceptions pierce through the veil of *maya;* we can stop at any point, and say, "This is real," and it *will be* real, because we have made it real. The limbs, that have long been amputated, feel real; the pain of these phantom limbs can sometimes be excruciating. The stumps of an arm or a leg, or even a uterus after hysterectomy, may have a lingering memory, and be an irrefutable presence. And the memory of a dead child, or a deceased beloved, may have real and unbreakable links. They are all illusions, yet they are real.

It is not only Alice who wanders in the wonderland of shifting perceptions, we all do. Alice says: "If I had a world of my own, everything would be nonsense. Nothing would be what it is, because everything would be what it isn't. And contrary wise, what is, it wouldn't be. And what it wouldn't be, it would. You see?"

For too long our sages and philosophers have talked about truth as though it were apart from the illusions and phantoms of life. Pain of the body, and the anguish of the heart, especially of others, were often dismissed as a mere illusion. For too long, in the elusive search for the Brahman, for 'ultimate reality,' we closed our eyes and ears, and above all our hearts, to the wonders and glories of, what we arrogantly believed, was only *maya.*

Hunger of the body is real; sobs of loneliness are real; the hurt caused by those who walk away with disdain, is real. So much that is so palpably real, we so arrogantly make unreal. We create a maze of words and theories that avoids the muck of life in all its grandeur, and instead formulate a carapace of glorious delusions. We embrace not new birth but rotting carcasses. "'He (who is) not busy being born," said Bob Dylan, "is busy dying."

And we have indeed been very busy dying.

Not infrequently, men of extraordinary genius – Pablo Picasso, Ernst Hemingway, Richard Wagner, Bertolt Brecht, Jean-Jacques Rousseau, and above all Isaac Newton – lack generosity of spirit; they soar in their art but not always in their lives. Great civilizations and grand philosophies too can be mired in arrogance and be self-serving. Truth then becomes only a phantom limb, only a memory that once existed, but is now source of real pain.

For truth, one can go to the words of Bhagavada Gita, or of Adi Shankara; often the weight of their erudition threatens the sweet bird of life. Instead, one may go to the simple, haunting and tantalizing lyrics of Joni Mitchell, and find a solace that is as illuminating as any truth:

I've looked at life from both sides now
From win and lose and still somehow
Its life's illusions I recall
I really don't' know life at all
I've looked at life from both sides now
From up and down, and still somehow
Its life's illusions I recall
I really don't know life at all.

As one gazes through the mists of time, so many images appear and disappear fleetingly. What is real? What has become real for me? What have I made real for my consolation or pride? What truth is hiding in those ghostly apparitions that wander all over me? What is it that keeps haunting me in these shadows?

I *re-member* and thus I hope to become!

❖

PART II

How's and Why's of the Universe

I refuse to accept despair as the final response to the ambiguities of history. I refuse to accept the idea that the "isness" of man's present nature makes him morally incapable of reaching up for the eternal 'oughtness' that forever confronts him. I refuse to accept the idea that man is mere flotsam and jetsam in the river of life, unable to influence the unfolding events which surround him.

— Martin Luther King

7

The Eye of the Beholder

In 1564 were born two men of extraordinary genius: one in England and the other in Italy. Both of them – William Shakespeare and Galileo Galilei – opened our eyes to a different and vaster world; one probed a world that is woven by a thousand strands of imagination, thought and action, much of it within the human mind itself. And the other set his gaze in the outer sky, and farther and farther, at the moon, and the planets. Both, reaching into the human soul for answers to the perennial 'why', and reaching out to nature 'out there' for the 'how', have been essential to the human journey, despite all their perils and trepidations. It was so four hundred years ago, as it was two thousand years earlier, and it is no less today. One might perhaps even suggest that the 'why' and the 'how' of our seeking constitute a sort of 'Double Helix'; they are interwoven; together they carry the secret of life. Thus, whatever ladder we have tried to climb to any paradise, however earthly, however ethereal, has always rested on these two rungs.

The human eye, even as it gazes outside – often aided by powerful instruments created by our scientific ingenuity – is also forever looking within; indeed, as it witnesses the sumptuous fecundity of the universe in all its grandeur and complexity, it ceaselessly envisions that universe anew. It is not only 'what is' that enraptures it; it is also 'what ought to be' that enshrines its insights and foresights in the 'mind's eye'.

It is thus that ethics and its hundred different facets – justice, freedom, honour, dignity, meaning and purpose, rights and duties, *dharma* and *karma* – assert themselves, and draw their rationale and inspiration from religions and cultures, from myths and literature, from customs and habits, from dialogue and debate, and from the vicissitudes of life itself. In turn, they create dreams and visions of human destiny, however flawed, however ephemeral.

In 1897, French artist Paul Gauguin created a large painting in Tahiti, considered by many as his great masterpiece, entitled: *Where Do We Come From? What Are We? Where Are We Going?* It is a haunting work of exquisite beauty with brooding images of life and its inevitable decay. Gauguin created this painting in a remote island, possibly quite unaware of, or perhaps oblivious to, the debates and discussions swirling around the publication in 1859 of Charles Darwin's great book *On the Origin of Species* that set out to establish the scientific foundation for the origin and evolution of life. Possibly unaware, even of the publication of *On Liberty* by John Stuart Mill the same year, all at the height of European imperialism.

In our age of science and political liberalism, one may wonder, is there any place for Gauguin and his soulful enquiry?

With steadfastness that borders on evangelical fundamentalism, there are some, like Richard Dawkins, who insist that the only true path to human salvation is through science. "Darwin provides a solution, the only feasible one so far suggested, to the deep problems of our existence," Dawkins proclaims. "We no longer have to resort to superstition when faced with deep problems: Is there a meaning to life? What are we for? What is man?"

After posing the last three questions, Dawkins tells us, the eminent zoologist G.G. Simpson concluded that all attempts to answer them before 1859 – when *On the Origin of Species* was published - were 'worthless' and that "we will be better off if we ignore them completely."

Though a revolutionary thinker, Darwin was, nevertheless, very much part of his times. As a Victorian gentleman, he donated to the South American Missionary Society for turning around the 'savages' he encountered in Tierra del Fuego, and asked to be made an honorary member of the Society.

Yet in the name of Darwin so much biological fundamentalism has been spread for decades: 'Survival of the Fittest' led to 'Social Darwinism', and to 'natural' justification for 19th century European imperialism, and to world-wide eugenics movement, most infamously in Nazi Germany, where the weak and the feeble-minded were eliminated by hundreds of thousands for the 'betterment' of the human race. In 1840s, as the Irish were dying in hordes because of the potato famine, many in England were convinced that any attempt to help them would be to interfere in the 'laws of nature'. Indeed, the head of the English Treasury had remarked so casually in 1848, "To feed or clothe the dying would be to interfere with the free market."

It is one of the great ironies in the history of science that the man whose ideas and theories were abused for rationalizing such misguided notions was a man of extraordinary compassion and sympathy, though not always without the prejudices of his age or his class. Darwin wrote unambiguously about ethics that must guide human conduct, and how natural selection seemed to no longer act upon civilized communities in the way it did upon other animals:

> With savages, the weak in body or mind are soon eliminated; and those that survive commonly exhibit a vigorous state of health. We civilised men, on the other hand, do our utmost to check the process of elimination; we build asylums for the imbecile, the maimed, and the sick; we institute poor-laws; and our medical men exert their utmost skill to save the life of every one to the last moment. There is reason to believe that vaccination has preserved thousands, who from a weak constitution would formerly have succumbed to small-pox. Thus the weak members of civilised societies propagate their kind. No one who has attended to the breeding of domestic animals will doubt that this must be highly injurious to the race of man.

> The aid we feel impelled to give to the helpless is mainly an incidental result of the instinct of sympathy, which was originally acquired as part of the social instincts, but subsequently rendered more tender and more widely diffused. Nor could we check our sympathy, even at the urging of hard reason, without deterioration in the noblest part of our nature.

Darwin called sympathy "the noblest part of our nature", and that it must not be checked "even at the urging of hard reason." It was not a 'scientific' observation, based on any empirical data. It was part of Darwin's dream of what he considered to be good and noble life. This is when, one might say, science surrenders to spirit.

Sympathy. Empathy. Love. Altruism. For a long time we have struggled to see the face of our brother as our own. Who am I, and who is that other in the mirror or across the river? Reflections on life and its pangs, on death and its inevitability, and on meaning and purpose of life that Gauguin so poignantly portrayed in his painting have been the warp and weft of literature and the arts for millennia, both, before and after 1859; they are also intricately and subtly woven in our myths and folktales. Great achievements in science and medicine over the past several hundred years have not lessened the poignancy of these questions; they stare us in the face in the early years of the twenty-first century with the same urgency as they may have done at the dawn of civilization, ten thousand years ago.

"The truth of life" – if such a phrase can be used in the age of science – must be a many-splendored phenomenon; it is illuminated, grasped and beheld by myriad sights and insights, of poets and sages no less than of molecular biologists; if a mathematical equation stands forever, as Einstein believed, words uttered a thousand years ago by a sage can also haunt us like some cosmic force for centuries to come.

Four hundred years ago, in 1609 in Venice, as Galileo was beholding the 'beautiful and delightful' body of the moon through a telescope, he complained bitterly against "the principal professor of philosophy who I have repeatedly and urgently requested to look at the moon through my glass [telescope], which he pertinaciously refuses to do."

Whether for professors of philosophy, or savants of science, there remains a cosmic chasm between us: the challenge of looking at the moon through the eyes of another. This may indeed be the most demanding expression of sympathy and "the noblest part of our nature."

It is not a lesson one can learn in a laboratory, nor by peering at the farthest reaches of the galaxies through a radio telescope. It is a lesson we forge in the smithy of our own soul, in the heat of the long night.

❖

8

A Flower in the Volcano

"If your heart is a volcano," wrote Kahlil Gibran, the author of *The Prophet*, "how shall you expect flowers to bloom?" In the age of science many find it difficult to believe that our heart or our inner eye is what so often defines, confines and refines our universe. Is it possible that when our heart is in turmoil and this eye is distorted, or unfocused, or enraged, a ruby out there, or a glorious sunset, or a sumptuous feast may mean little, or may even seem jarring?

There are eyes and hearts that find, however momentarily, a harmony, and a gentle beauty even in the most mundane. "I never saw an ugly thing in my life," eighteenth-century English artist John Constable once declared. "Let the form of an object be what it may – light, shade, and perspective will always make it beautiful."

One must ask: Is it possible that with a different perspective, in a different light, in the eyes of a true witness, all things become beautiful?

Beautiful and full of grace?

Is it possible?

There are at least two different streams of perception and knowledge that have blossomed hand-in-hand throughout the human journey: one probes the tangible and measurable forces that twist and turn the phenomenological world – the world of senses and experiences – and give it shape and form:

how do the petals of a rose get their texture and hues? How and why do the bees gather around a blooming flower? How do Monarch butterflies navigate their way over thousands of miles to reach a small isolated place in Mexico?

This is the grand and wonderful world of science; it is not a world merely of facts, but of facts re-interpreted, re-connected, re-examined, and re-visioned. This manner of knowing the natural world – the flowers, the rocks, the birds, the rivers, the distant stars, the genes and atoms – has been exploding over the decades and centuries, and has moved far beyond human senses and rationality; sometimes it has turned into a mere mathematical abstraction. Yet for all that, it has been a noble and exalting activity for millennia, and its pursuit has sometimes been very intoxicating. "Deny the powerful and their warriors entry to your workshops," warned the alchemists – the precursors of modern science. Greek geometrician Euclid, when asked by someone what was the practical use of his ideas, is reported to have told his slave sardonically: "He wants to profit from learning – give him a penny."

Pythagoras, the founder of Greek Mathematics in sixth century BC, offered a hundred oxen to the Muses in thanks for the inspiration for his great theorem named after him. To his followers, he was a magician and a mystical seer who saw harmony everywhere, in nature and in music, that could be expressed in numbers.

There is yet another mode of perception that too connects seemingly disparate things and events and draws new meanings from them, thus imbuing them with certain beauty and wonder. "If we could see the miracle of a single flower clearly," said the Buddha, "our whole life would change." This manner of seeing the flower is not that of science, but of poetry, of a sage, of a different eye. "To see the world in a grain of sand," invoked poet-mystic William Blake, "And Heaven in a Flower."

This mode of 'seeing' is as precious as that of science, and what it reveals is even more grand and ennobling. What science, for instance, could match the insights of poet Dylan Thomas in the following words and images?

The force that through the green fuse drives the flower
Drives my green age; that blasts the roots of trees
Is my destroyer.
And I am dumb to tell the crooked rose
My youth is bent by the same wintry fever.

It is this imaginative power that weaves the world of senses into something new, and something exalting; this power was acknowledged, for instance, for the 2011 Nobel Laureate in literature, Tomas Tranströmer, "because, through his condensed, translucent images, he gives us fresh access to reality." Without this power we would be greatly diminished in our being. This power is the fountainhead of our artistic and spiritual longing and its blossoming. And this power is different from science, not in opposition to it, or as a substitute for it, but in a realm of its own.

There is a story told about the pioneering physicist Ernest Rutherford at Cambridge University towards the end of the First World War. He once failed to attend a meeting of the British committee of experts appointed to advise on new systems of defence against the enemy submarines. When he was censured for his absence, he retorted without embarrassment: "Talk softly, please. I have been engaged in experiments that suggest that the atom can be artificially disintegrated. If it is true, it is of far greater importance than a war."

Wars and dictators come and go, said Albert Einstein, but "a mathematical equation stands forever." The same could be said about a poem that has travelled the memory lanes of a forlorn lover over centuries, and still evokes warm tears of some primordial longing: *gar main jaanti, prem karyo dukh ghanero hoye...* Had I known the pangs of love, I would have declared from door to door, Love Not. Love Not."

Science, it is sometimes said, set out to read the Book of Nature as people have read the Holy Book or the Book of Revelations. Yet every student of literature knows that no book, of any worth, can be read literally. One has to learn to read between the lines; one has to search for allusions and metaphors, for parables and allegories; one has to listen to the barely articulated whispers; one has to be prepared to hear the distant thunder. No book, however great, can contain all the murmurs of a throbbing heart, or all the sobs and sighs of thwarted dreams. It can be, even at its best, only a finger that points at the moon. It is not the moon.

It is thus that our investigation and study of nature 'out there' cannot be only 'scientific'. Nature is a great treasure house of metaphors which reveals its true grandeur only in the imagination of a poet or of an artist or a sage. "When a flower blooms," said Sri Ramakrishna, "the bees come uninvited."

In such simple words, the great seer rendered a new meaning to the power of gravitation between hearts and minds, a meaning that is different – and so scintillating – from Newton's Law of Gravitation.

There is a story told about a young scientist in Los Alamos, where the first Atom Bombs were developed during the Second World War. Walking down the street one evening, the budding genius of science was observed to be bearing a smile of almost angelic beauty. He looked as though his inner gaze was fixed on a world of harmonies. He was, in fact, he said later, thinking about a new mathematical problem whose solution was essential to the construction of a new kind of hydrogen bomb.

For this scientist, as for many others, research in nuclear weapons was yet another problem in pure mathematics, unsullied by cries of children or destruction of civilizations. All that, he is reported to have said, was none of his business. He had even refused to watch trial explosions of any of the bombs that he had helped design. Nor did he wish to visit the cities of Hiroshima and Nagasaki, though he was invited; or even look at the pictures of destruction wrought by the two bombs.

This great scientist was quite possibly at peace with himself. Or equally, he may have been at war within, or with his father or his youth; he never did tell us who he really was. Relentless pursuit of mathematics or physics may even have been a masquerade for him. Or some form of sublimation. People in turmoil can still create great science, or extraordinary music, or remarkable paintings. We should look only at their creations, we are sometimes told, and not at the creators. But are the two so distinct from each other?

In her novel, 'Unless', Carol Shields wonders what the Theory of Relativity means to the life of a physicist recently separated from his wife. "Would you say," he is asked, "that the theory of relativity has reduced the weight of goodness and depravity in the world."

"Relativity has no moral position," he answers. "None whatever."

Without a moral position, Relativity – and Science – cannot know the world of the living and of the conscious. There is much it can know, and has known, but there is also much that is beyond its compass. In his office at the Institute of Advanced Study in Princeton, Albert Einstein had a sign hanging that read: "Not everything that *counts* can be counted, and not everything that can be counted *counts*."

It is the great achievement of science that so much natural phenomenon has been measured and counted, and has thus been made objective and real. It is, however, the formidable limitation of science that it refuses to accept that so much that counts in life can hardly be counted. What counts, and what *should* count, are questions so inseparably wrapped around the human condition; the answers to these eternal questions are not to be found in science but elsewhere.

For, it is not only where we look, but how we look that counts. "Eyes and ears are bad witnesses for men," says Greek philosopher Heraclitus, "if they have barbarian souls." Without compassion, and without an ever-expanding circle of compassion, however far we explore the outermost galaxies in the cosmos, we remain shackled by the blinders of our own perceptions.

"The world will never starve for want of wonders," wrote G. K. Chesterton, "but only for want of wonder." A sense of wonder that renders meaning and insight to our eyes and to our being was no less in the man who saw his reflection in the water for the first time than in the man who looked at the far galaxies through the Hubble telescope. That great moment – so steeped in chaos and enigma – when the universe is said to have burst forth, in fact, occurs again and again, every time a seed bursts, a life comes into being, or an idea takes root. That's how a person, even though completely ignorant of the facts and theories of science, can and does partake in the wonder that is the universe.

The inner eye is our greatest gift before which all other instruments pale immeasurably. "Cleanse thy doors of perception," William Blake implored in the nineteenth century – as sages have done for millennia – "and you will see things as they are: Infinite."

For this inner eye to *see*, often we have to close other eyes; certainly we have to be at peace with ourselves. Without this peace – without inner calm, without mindfulness, without serenity and compassion, without an unconditional acceptance of the universe as it is – as in a turbulent river, we cannot see a clear reflection of ourselves. Or of our neighbour. Or indeed of the universe at large.

In that turbulent state, we remain somewhat blind. Obstinate and blind. Steeped in despair or cynicism. Without a touch of hope. We may ceaselessly go on exploring the farthest edges of the galaxies, and yet may be paralyzed from taking the very first step across the cosmic chasm that separates us from ourselves.

Think of it: "If seeds in the black earth can turn into such beautiful roses, what might not the heart of man become in its long journey toward the stars."

❖

9

The Tides and the Moon

The great composer Ludwig van Beethoven died in 1827 in the midst of a thunderstorm. It is said that his last words were: "Plaudite, amici, comedia finita est ... Applaud, my friends, the comedy is over." For a man who was almost completely deaf in the closing years of his life, any call for applause must have been a rather ironic gesture.

During much of his tumultuous life as a composer, Beethoven had insisted that "music is a higher revelation than all of wisdom and philosophy"; this at a time when religious thought was still well-entrenched, and the Book of Nature was being read by natural philosophers with greater fervour than ever before. At his funeral over twenty thousand mourners lined the streets of Vienna. Among the mourners were musicians, poets, princes, and all those men and women who had been 'set on fire' by his music.

Beethoven was a great musician, of course, but he also had the daring to tell an age, increasingly touched by a new wave of Enlightenment, about the purpose and meaning of life. He reminded the princes often: "What you are, you are by accident of birth; what I am, I am by myself. There are and will be a thousand princes; there is only one Beethoven." It was thus that at his funeral, as thousands wept, composer Franz Schubert spoke for all lovers of music: "We stand weeping over the broken strings of an instrument now stilled."

That the great composer was like an instrument of music himself, out of which flowed music of enchanting beauty, makes one wonder if, in some mysterious ways, the creator is not so very close to the Creator. Music, the musician, and the instrument of music are a form of trinity; at some stage, I wonder if they don't coalesce into one.

At the age of twenty-one, when I had just arrived in Canada to study physics, I had a glimpse of an uncanny proximity between them. And I was dazzled.

It was a Sunday afternoon in early September; I was away from home for the first time, and was now staying in a rooming house on Sussex Avenue, on my own, with a bunch of strangers.

Suddenly from the adjoining room came sounds of music I had never heard before. I was transfixed; I could not believe that such music could be created by any human hands or lips or imagination. As a frail young man, despite all the pride of a budding intellectual, I found myself shaking like a leaf. Is it possible, I wondered, to be in the presence of such grandeur, such unfathomable beauty, so near, so overwhelming, and yet so utterly unknown to me for so long.

A small sliver of afternoon autumn light was sliding through the narrow window in my tiny room, yet I felt the whole universe was bursting with illumination. This was a gift – unasked, uncalled for, thoroughly, utterly undeserving. I knew hardly anything at all about music; certainly I had no acquaintance with the music of the West. I had studied the theories and insights of great physicists and mathematicians from Germany, England, France and Denmark who had transformed our vision of the universe in the twentieth century. But this – as tears trickled down my cheeks uncontrollably – this was a novel and transforming experience. I knew how the distant moon, a quarter million miles away, could cause the ebb and flow of tides here on earth. But how some sound waves could bring heaven on earth, I had no idea.

When this ineffable music stopped, after what felt like an eternity, I rushed over next door: "What music were you playing?" I gushed at the stranger with the gramophone. "Don't you know?" he asked incredulously. "This is Beethoven. This is his Ninth Symphony."

Beethoven and his creeping deafness. Ninth Symphony. Ode to Joy, 'a hymn to the unity and freedom of humanity'. Eighteenth century European music. No, I didn't know. I didn't know at all. Yet I knew that these sounds had touched me far beyond anything I remembered, or could imagine. I felt grateful at this discovery. I felt blessed.

Some two hundred years after Beethoven's death, in June 2009, when the great sarod player Ali Akbar Khan passed away in California, he is quoted to have often remarked to his students: "Sound is God - *Nada Brahma*", and he insisted that this sound be created with certain veneration, not haphazardly, not heedlessly: "If you practice for ten years, you may begin to please yourself," he would say. "After twenty years you may become a performer and please an audience; after thirty years you may even please your guru. But you must practice for many more years before you finally become a true artist ... then you may please even God."

Please God! How very absurd that seems to so many of us now, the children of Enlightenment. Art for Art's sake. Art as self expression. Art as propaganda. Art as commerce. Art for flattery and vanity. Art for sensationalism. "Art is whatever you can get away with," declared pop artist Andy Warhol; indeed he got away with much that was far from art. Despite this all, perhaps some among us still cannot cease to search for the moon that can raise a roaring tide here on earth, in the marrow of our bones.

Karen Blixen, Danish writer known for her book *Out of Africa*, once advised a young writer, still struggling to find her tongue, "to write as though you owe an answer to God." How much closer would an artist be to the fountainhead of creation, if she is thus called upon to answer for all one's distractions and cacophony of sounds?

It is thus, in the Danish film *Babette's Feast* (1987), based on one of Blixen's short stories, on a solitary, windswept island, amongst spiritually parched but religiously dogmatic puritans, a great chef, Babette, presents a feast for a group of people who had forever subdued their taste buds for the mistaken glory of God. Babette's feast is no ordinary feast; it is a feast of celebration of grace that so often eludes those who are in denial of life, in all its overflowing fecundity. The feast is Babette's answer to God, and she stakes her all on it. Her entire earthly treasure, that could have given her freedom to return to the world of fame and adoration in Paris for her great culinary skills, has been squandered over those who know nothing of pleasure or joy. Yet she has chosen, like a Bodhisattva, to struggle on to bring a glimpse of a sensuous, erotic hope to those who have none. "Now you will be poor the rest of your life," they chide her, "like us." To which Babette replies: "An artist is never poor."

The artists are, of course, always poor, but how poor are they really? When anyone among us shows the 'courage to be', to leap across the chasm of fear and uncertainty, and dances to the echoes of '*Jai Ho! Jai Ho*... Ode to Joy... *Freude! Freude!*' we are invited to the most sumptuous feast at the great banquet of life. We then become the musician, the instrument of music, and the music itself. We become artists, creators and sustainers of life; we become Brahma and Vishnu, and we walk hand in hand with the Creator. We are rich. We are very rich indeed.

Each one of us is capable of being endowed with such abundant wealth; in the words of art-philosopher, Ananda Coomaraswamy, "an artist is not a special kind of person but rather every person is a special kind of artist."

One can walk on many roads, fly over many continents, and cross many oceans, and yet go nowhere; one wanders on the earth in poverty. To bask in the glow of abundance, a man must feel the pull of the moon, so he can behold the soaring tides in the ocean of his own being.

Many years ago, on a Sunday afternoon, in a new country, far away from the chatter of thoughts of science or philosophy, oblivious to the anxieties of my youth and of my future, against all odds, in defiance of all laws of probability and statistics, out of nowhere, in the company of a stranger, in a tiny place, on a cheap gramophone, a new delicious world was opened to me fleetingly. It is as though, on a windswept island of my parched soul, Babette had laid a feast just for me.

For a moment I thought I should write a letter to my unlettered mother in the distant and eternal country I had left behind. Should I tell her, I wondered, that I was at a banquet, and that this food tasted a little different from what I had eaten so contentedly at home?

❖

10

When Stones Shed Tears

How often is a city called a city of "poets, wine, women, nightingales and flowers"? Once I was in one such city, in the ancient and great civilization of Persia, in the Iran of today. This city of Shiraz goes back into antiquity; it was here that the oldest sample of wine, dating back to some 7,000 years, was discovered in clay jars; it was here that two of world's greatest poets – Hāfez-e Shīrāzi and Abû *Muslin bib* Abdallâh Shîrâzî, better known by his pen-name as Saadî – lived and died.

One summer day in early May, I was at *Hafezyieh*, the shrine of poet Hâfez; he was buried here more than six hundred years earlier, in 1390. If the Greeks loved their dramatists, the Persians loved their poets, more ardently than perhaps any other people in the world. Even today Hâfez's *Divan* is to be found at the home of most Iranians who recite his poems with a lover's passion, and use his words like proverbs, and a panacea for anything that ails the heart. Hâfez's poems are lyrical, they are ecstatic and mystical, and they are replete with allusions; they are rich and they are varied. And they are incorrigibly full of love and longing. For whom? One can hardly say; his longing is a great mystery, and it is utterly enchanting. It touched Goethe in Germany, Emerson in America, William Jones in England. And on this summer day in May, it touched a young woman fully cloaked in black. She kneeled at the shrine, recited some poems of Hâfez, and then cried, and cried, and cried.

It was a sight to behold, like a thick black cloud hovering over a snow-covered mountain peak, pouring its heart out.

How dull the world would be without romance and mystery? Without tears of joy and longing? Without the sense of the ephemeral and the unknowable? Without dreams quietly thwarted? Without questions, with no answers?

When the heart aches without knowing why; when it rejoices without rhyme or reason; when sorrow comes without a shade of darkness; when happiness just drifts by like an autumn cloud, when music shoots forth from somewhere, and words assume a mythopoeic reference, perhaps then a poem is born. And it travels hundreds of years down a memory lane to touch the heart of a forlorn lover in a forbidden country. Time does not matter then, nor context or truth; it has a transcendent meaning of its own.

In the heart of Paris there is a large cemetery called *Cimetière du Montpamasse*; here some of the most famous philosophers, writers, poets and artists of France are buried. Here is Jean-Paul Sartre in one corner, Samuel Beckett in another, and Jean-Pierre Rampal in another. One summer on my visit here I saw a young woman standing in silence before a stone carving, and place a flower on the grave. And then slowly a stream of hot tears began to trickle down her cheeks. I stood at a distance transfixed by this special moment, not knowing whose grave stone she was standing by. What line of the poet, what thought of the philosopher, what picture of the artist long passed away had touched her heart and how, I didn't know. Yet it was obvious that she had been moved to tears by a certain memory, a reference and a connection that was uniquely her own.

After she had moved away enveloped in melancholy and tears, I sidled up to the grave stone to read the inscription. It was Charles Baudelaire, a 19th century poet, whose name has become, over the decades, a byword for literary and artistic decadence, but many of his works, particularly *Les Fleurs du mal*, have come to be regarded as classics of French literature.

There must be, I thought, a certain solace to the human imagination to place the present in an on-going stream of time, to find an uninterrupted continuity in life. Whether six hundred years or six thousand, with all the changing faces of life, one longs for that eternal, immortal moment when things are the way they have always been. Then one basks in that discovery and rejoices, despite all the tears. A poem then, a myth or a painting, reverberates afresh; it becomes an anchor to which we tie our boat of life, and feel serene for a moment. Without such an anchor we are somewhat adrift in a sea of emotions, imagining our lives to be so unique that we become either arrogant or lonesome. To see our life forever lived, and living, is to savour its continuity, as one mighty river of life.

At whose grave, or *samadhi* I would be so moved, I sometimes wonder. When the great rebel sage-poet Kabir died, it is said, that both Muslims and Hindus argued and fought over his dead body, claiming him to be one of their own, and thus insisting on conducting the last rituals, of burial or cremation, accordingly. Throughout his rebellious and illuminating life, Kabir had argued that he was neither a Muslim nor a Hindu but a "creation of five elements in which light weaves its play." In his songs and *sakhis,* Kabir had indeed woven such images that they seem like a tapestry of illumination: "laali mere laal ki, jit dekhu tit laal... The light of my beloved is permeated everywhere. As I went searching for it, I too became part of the light."

The legend has it that as the Muslims and the Hindus fought over the body of Kabir, when the shroud was removed, there was no body to be found, but only a heap of flowers. The Muslims, it is said, buried the flowers and the Hindus cremated them.

Sophocles' Greek tragedy *Antigone* was first produced some 2500 years ago, in 441 BC; it has been called as "a work of art nearer to perfection than any other produced by the human spirit." Over the millennia, the characters of Antigone and Creon, the king, have become, respectively, archetypal symbols of resistance and of oppression. As such, the play and its mythos have inspired numerous versions and interpretations all over the world, and have served as a fierce rallying cry against wanton political brutality everywhere: in Nazi Germany, Mexico, Poland, India, Albania, South Africa and in modern Greece itself; in a period of one hundred years, between 1699 and 1800, more than 30 operas on the theme of Antigone were produced.

Representative of a fierce debate between 'moral idealism' and 'political realism', since the 5th century BCE, the *mythos* of this archetypal tragedy continues to reverberate to this day in far off lands, cultures and times; one wonders what masks and faces – both literal and figurative – have been given to *Antigone* by its various interpreters, and why.

In 1960s, when Nelson Mandela was a captive in South Africa's notorious Robben Island prison, the inmates presented a version of Antigone at Christmas. Mandela himself chose to play the role of the tyrant king Creon, who, before he reluctantly assumed the kingship after the death of his brother Oedipus, was a scholar, and a lover of the finer things in life. Now unexpectedly thrust into power, he has to be a 'political realist', balancing the duties to the state with the affections of the heart. As Creon, Mandela is said to have uttered the following words, that came to assume a very special significance when, against all odds, he ushered in a new non-racial democracy in South Africa and became its first black President:

> *Of course you cannot know a man completely,*
> *his character, his principles, sense of judgment*
> *not till he's shown his colour, ruling the people,*
> *making laws. Experience, there's the test.*

We are all tested by our actions. Yet every little step we take seems to have been guided by some hidden hand, some thoughts perhaps long forgotten, some intuitions, some memories. A faint beckoning from the words of a poem, a myth, or a story long heard, lingers on in our being. It takes hold of us, it digs deeper and makes roots, and then it begins to burst forth. That is how, for me, with my roots in India, the *Ramayana* and the *Mahabharata* are the fountainheads of my existential battles; their allusions and metaphors still nourish my imagination. They are the ground of my being from which so much shoots forth that is far from my choosing, and I wonder if what we call 'free will' is all that free.

That's how it must have been for that fully cloaked young woman in Shiraz, in the city of poets, nightingales and wine, before the shrine of the poet Hâfez; that's how it must be for the French woman in Paris before the grave stone of Charles Baudelaire. And that's how Nelson Mandela must have drawn inspiration from the Greek tragedy *Antigone* first presented some 2500 years earlier.

❖

11

Who are the People of the Book?

In 1543 were published two very remarkable books in Europe that can be said to have ushered the Scientific Revolution: one was *De revotionibus orbium coelstium* ('On the Revolution of Heavenly Bodies') by Polish astronomer Nicolaus Copernicus, and the other was *De humani corporis fabrica libri septem* ('On the Fabric of the Human Body in Seven Books') by Andreas Vesalius. In their subject matter and exploration, the two books were as different as they could be; one questioned the place of the earth in the solar system, and the other probed the subtle and complex structures of our own bodies.

Great ideas need great vehicles to be transmitted and to spread. Before the scientific revolution could take root as a new culture of open and uninhibited enquiry, its essential principles needed to be known, examined, understood, debated, and then perhaps assimilated in the culture of a people. The invention of the first movable type printing press a hundred years earlier in Germany proved to be this great vehicle; it led, within a few decades, to an explosion of book printing all over Europe. By 1500, the printing presses in Western Europe had already printed more than twenty million volumes. In another century, the number was to grow ten times, to some 200 million copies.

These books were not merely religious in content; they were about all kinds of subjects – political, scientific, medical, philosophical, and literary. Five hundred years ago, fifteenth century scholar, Erasmus, known as the 'Prince of the Humanists', was a 'best-seller'; at least 750,000 copies of his books were sold during his lifetime alone (1466-1536). His books included such titles as *In Praise of Folly* and *On Civility in Children.* Already, even before Columbus set sail in 1492 to find a westerly route to India, there was a sense that the metaphysical foundations of the world were shifting.

The revolutionary potential of bulk printing took both the European princes and the papacy by surprise. If Luther's tracts could be distributed in 300,000 printed copies, and light a zeal for Reformation, Charles Darwin's *On the Origin of Species* (1859) could equally compete for the hearts and minds of men about the origin and evolution of life on earth.

However, unlike listening to sermons and lectures, or later to radio, or watching television and films, faculty of reading needed to be cultivated; one could not simply read as one could listen. Literacy has been a long time coming, and its spread became an essential public goal in the new age of Enlightenment that was slowly dawning. To be 'illiterate' now became synonymous with being ignorant. As late as 1840s, about forty percent of men and women in England could not sign their name.

There is a vast chasm between the oral and the written world. Though there have been rich oral traditions of communication in many societies, such as India and Persia, and where great tracts of religious rituals and incantations in verse-form have been passed down from one generation to another, oral communication is endlessly vulnerable to distortions and whims. It is the book that has the power to command awe, respect, knowledge, insights, dreams, and above all to initiate debate, discussion and interpretation, and even fury. In one way or another, we are all people of the book; that is why, I believe, all revolutions – scientific, political, religious, and social – have a book that stirs passion and a following. Where would Marxism be without *Das Kapital*? It appeared in Russia in translation in 1872 before there was a

Russian translation of the Bible. The Russian censors were convinced "that very few people in Russia will read it and fewer will understand it." The book's first print, however, was sold out within the year, and Marx felt certain that it was there that the book "was read and valued more than anywhere else."

Every religion has its 'holy' books. For the Sikhs, for instance, the Grantha Sahib is the eternal guru. The Bible, the Koran, the Torah, the Vedas and the Bhagavada Gita are sacred; for the believers, they are words of God: infallible, beyond debate, immortal, eternal. Non-religious, secular literature too has been crucial to social change: *Uncle Tom's Cabin: Life Among the Lowly*, written by abolitionist Harriet Beecher Stowe in 1852, is said to have laid the groundwork for the end of slavery in USA. Arguably, Charles Dickens' novels were no less significant than the theories of his contemporaries, Karl Marx and Charles Darwin, in stirring a new revolution in human thinking.

In 1952, a 54-volume set of 'Great Books of the Western World' was released; it was hailed as 'an act of piety'. "Here are the sources of our being," it was declared. "Here is our heritage. This is the West. This is its meaning for mankind."

Books have inspired, enlightened, educated, stirred, amused and informed young and old everywhere. Yet some of the books have been reviled; they have been banned; they have been ceremoniously burned; they have been condemned out of moral indignation, hatred or ideological or racial purity. The great Library of Alexandria was burned; the Library of Baghdad was obliterated; under China's Qin Dynasty, books were burned and scholars were buried; the Spanish conquistadors and priests destroyed Aztec codices and the Nazis burnt Jewish literature. In 1950s in the USA a number of books by psychologist William Reich were ordered to be burnt under judicial orders.

During the Spanish Inquisition, the Koran was burnt ritualistically. In his 1821 play, *Almansor*, the German playwright Heinrich Heine refers to such burnings: "Where they burn books, so too will they in the end burn human beings." ("*Dort, wo man Bücher verbrennt, verbrennt man auch am Ende Menschen.*") A hundred years later Heine's books were among the thousands of books that were put to flame by the Nazis in Berlin's Opernplatz.

In 1873, "New York Society for the Suppression of Vice" was founded, and book burning was declared its worthy goal on its seal; the Society is believed to have burned more than 15 tons of books and 400,000 pictures.

Yet for a thousand acts of wanton destruction, suddenly there is an act of such intense love for a book that one feels deeply moved by it. During the Yom Kippur War in 1973, a young Israeli soldier so loved a book that he trimmed it on three sides so it could fit into the pocket of his uniform. One night, he told me many years later, in his army camp, against strict orders, he pulled out his beloved book from his pocket and felt impelled to read it. In the faint illumination of his flash light, he started reading:

Where the mind is without fear and the head is held high;
Where knowledge is free;
Where the world has not been broken up into fragments by narrow domestic walls;
Where words come out from the depth of truth;
Where tireless striving stretches its arms towards perfection;
Where the clear stream of reason has not lost its way into the dreary desert sand of dead habit;
Where the mind is led forward by thee into ever-widening thought and action—
Into that heaven of freedom, my Father, let my country awake!

The book was Rabindranath Tagore's *Gitanjali*.

❖

12

How Words Turn on Their Head

In a reading of a remarkable novel, *Sea of Poppies*, by Amitav Ghosh, I was overwhelmed by a most imaginative recapturing of the myriad ways a whole range of people spoke English and several other languages and dialects in 1830s in the villages and cities, and on boats and ships in India. The novel tells the story, at once shattering and illuminating, about the cultivation of poppies for the production of opium by the British East India Company, and its disastrous impact on the lives of tens of millions of people in India and in China.

In the hands of an imaginative writer, words of a language are like colours in the palette of an artist; he uses them to transform familiar landscapes into something lyrical, something haunting. If under a microscope we see a new world that is hidden from our eyes, a work of great literature reveals a world that otherwise just passes us by. It is as though something has been made real for the first time.

In today's India one is overwhelmed by the diversity of languages, dialects, accents and use of metaphors and allusions. There are some twenty-two major languages in the country and possibly a few thousand dialects representing different regions, castes, classes and professions. And people seem to speak endlessly, chattering about this and that, and that and that. A language is a vehicle for communication, but what is it that anyone is trying to communicate?

We gossip, chatter, inform, cajole, seduce, reprimand, inspire, explore, invoke, refute, argue, threaten, recite, sing, grumble, and on and on. The list is quite literally inexhaustible. Our personality, our self, indeed our being are wrapped in language. Without an adequate language, we feel shrivelled, hesitant, and incommunicative. We feel less of a human being.

Language gives us identity. It gives ways to laugh and play with words. It gives us poetry, a way to express thought and a feeling; it even teaches us how to go beyond words; it lets us imagine and build castles in the air.

We may be forgiven if we sometimes believe that language is primarily a human attribute; that it distinguishes us so clearly and so unequivocally from the rest of the animal kingdom. So much so that when news breaks out that a chimp or a baboon can recognize a few words or symbols, or comprehends some verbal commands, we feel so utterly in awe of their achievements. Yet if a human being fumbles for some words out of shyness or nervousness, or is tongue-tied or is feeling flustered, or mixes some words or phrases, we deride him and do not hesitate to declare that such a person is an idiot.

In truth, however, even such an 'idiot' is a veritable genius; how words and sentences are formulated, how they are arranged into grammatical coherence, how they refer to things – seen and unseen – to the past and to the times still to be, are endlessly fascinating.

"Where were the words before they were uttered?" "Where were the thoughts before they found the words to express themselves?"

I recall how once my two-and-half year old daughter Amrita so utterly surprised me. Seeing a candy in her hand, I implored her: "Amrita, candy no good." Without a touch of hesitation, she replied: "Papa, candy so good." I was amazed, how such a young child could use the word 'so' and with such imagination. In a state of awe, I said again: "Candy no good, no good, no good." As though she could see the rhythm in the words, without missing a beat, she responded: "Candy so good, so good, so good."

The acquisition of a language by a growing child is one of the great achievements of maturation; this achievement, possibly more than anything else makes humans stand apart from other animals. Acquisition of a language – any language – is indeed the greatest and the most splendid feature of us humans, and of our humanity.

Language has been called an instinct in man; this means the capacity to learn a language is bred in the human bone, as it were. This capacity is to be found in every human child, and this capacity is the most distinguishing feature of what it is to be human. This is the great gift of nature to man. Yet what language – or even languages – a child would actually speak would depend on what language(s) is spoken to her. Possibly there is no grander example in all of the animal kingdom where 'nature' and 'nurture' so profoundly and intricately act together to create the very essence of a person.

When a language, or a dialect, is spoken only by a few, its communicative power is limited. Each small group then has its own language; there may in fact be few incentives to reach out and touch another group. When the Europeans first arrived in Canada some 350 years ago, there were more than 70 languages spoken by the indigenous people. Each tribe had its own language. In order to create a nation, the French and the English forced their own language, particularly on the young, in segregated residential schools. The mouths of the children were washed with soap if they ever used their own languages, and they were often beaten.

All imperial powers – Romans, British, French, Spanish, Portuguese, Dutch – imposed their languages on the ruled, thus forging a new and alien identity. Canada is now struggling to cope with the two founding European languages, English and French; it is not very often that second or third generation Canadians speak more than one language. In India, as in many countries in Europe, on the other hand, it is so common to see even young children speaking two or three languages; this renders, I believe, a richer and more creative soil for life and culture to flourish.

Language as a basis for cultural, social and ethnic identity is a new and world-wide phenomenon, as more and more people, even as they reach out to touch a much larger world, assert their independence from any dominant group. The opposition to learning Hindi in Tamil Nadu came out of that sense of imposition. The denial to Kurds the use of their own language by Turkey became an international issue.

In the past three decades, English has certainly become the most important international language. The new globalized world demanded one common language for communication. Now there are more people in China learning English than the entire population of the USA, of 300 million people. One could say that English no longer belongs to the English, or to the Americans; it belongs to all those – in India, or China, or Japan or Malaysia – who use it, with their own flair and in their own accents and idioms.

Nevertheless, however complex a language may be, in its structure and grammar, one needs to remember that it was not created by a mathematician. A language evolves under a myriad social, cultural and political forces; it absorbs words and idioms from other languages; it coins many words and phrases of its own to reflect the changing world around it. And then the great and gifted practitioners of a language – the writers, poets, teachers, dramatists, and comedians – keep adding a million colours and hues into it to make it colourful, vivacious, vibrant, playful and ever new.

Starting from a two-year child who says, "candy so good" to Hamlet responding to Polonius: "What do you read, my lord?" with a terse "Words, words, words", there is no knowing how and when words may turn on their head, and someone may start speaking in tongues.

❖

Part III

Estrangement from the Universe

Then Peter went up to him and said,
"Lord, how often must I forgive my
brother if he wrongs me?
As often as seven times?"
Jesus answered,
"Not seven, I tell you, but seventy seven times."

— Matthew 18:21-22

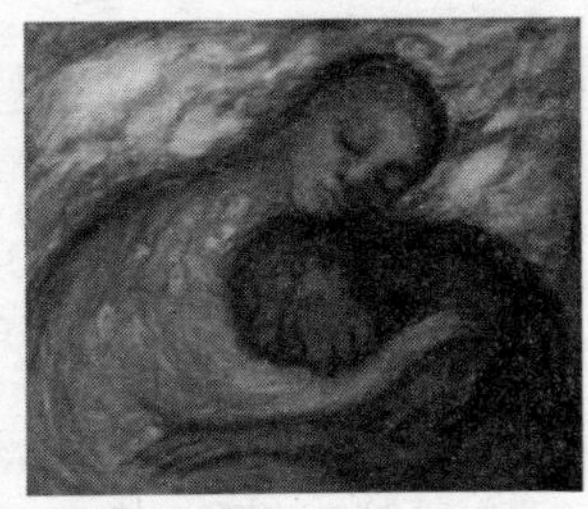

13

O Brother, Where Are You?

Garcia Lorca was a legendary poet; in 1936, during the Spanish Civil War, he was summarily executed by anti-communist death squads. His books were burnt, and no one knew where his body was buried. Over the decades, however, Lorca's poems have found an abiding place in the hearts of millions of people all over the world.

In the brutal Spanish Civil War, an estimated 500,000 people died, with both sides committing gruesome brutalities. In 2008, a Spanish judge, Baltasar Garzon, opened the first formal probe into murder and repression during the country's fascist era by ordering the immediate exhumation of 19 mass graves, including one that was believed to contain the remains of poet Lorca. The judge's order cited 114,266 people as missing or 'disappeared' under General Francisco Franco, declaring the repression and systemic extermination of political opponents during the Franco era as 'crimes against humanity'.

Stirring the past – revisiting it, re-membering it, acknowledging it, re-creating it – is never easy, whether the past relates to a family, or a community, or a nation. With each body that is exhumed there are a million slithering worms waiting; with each pyre that is turned over, there are a thousand sparks threatening to blaze once again.

Like the tongue that keeps stroking the aching tooth, for the sake of justice, or truth, or sheer revenge, or to 'set the record straight', we keep on exhuming the bodies, or turning the ashes over. One day it is the body of the Chilean president Salvador Allende exhumed to determine whether he committed suicide on September 11, 1973 as the military coup raged on, or whether he was murdered. On another day it is another body deep in some forest far off.

In our new globalized world, lies, half-lies or half-truth, historical evasions and national grand-standing are now increasingly coming under the public gaze. For over fifty years, for instance, all successive governments in Poland had persistently echoed the lies of their masters in Moscow about the 1940 massacre at Katyn in Poland where over 26,000 Polish officers were executed by the Soviets. The truth was finally established only in 1992, after the collapse of the Soviet Union, when secret documents were delivered to former president Lech Walesa by the Russians that proved conclusively that Stalin had ordered the execution.

The impact of this acknowledgement on Poland was enormous. Since the perpetrators had denied their guilt for so long and so persistently, a fierce bitterness had lain like a sword on Polish consciousness. Now, its public confirmation by the perpetrators had somehow made this agony, if not entirely vanish, at least bearable.

Could this be the great healing, or even redemptive, power of acknowledgement?

Yet, even as it was discovering one truth unequivocally, Poland itself was hiding another. In the summer of 1941, in a small town of Jedwabne, 1600 Jews were herded together in a barn and set on fire. For decades, it was said to have been done by the Nazis. Then it was revealed that the heinous act was committed not by strangers but by the anti-Semitic towns people themselves. In a remarkable play, *Our Class, by* Tadeusz Slobodzianek, one sees how a group of men and women, who had grown up together and had studied together in a school, stoop to betrayal and barbarism.

For so long the truth was hidden and distorted, and now, as one character in the play says it, "you can never bury the truth."

Memory is a double-edged sword: nations and individuals can suffer from amnesia, from remembering only very little of the past, or even denying it altogether. Or sometimes wallowing in the past, like a corroding addiction, and seeking revenge, passionately fuelling the fire of vengeance. Simon Wiesenthal, a survivor of Hitler's death camps, became the most dogged of Nazi hunters. He insisted that he was seeking 'justice not vengeance.' Yet it was said about Wiesenthal that the sight of SS officers in shackles filled him "with an elation akin to that evoked by divine worship."

Then there are times, when for some the memory of the past is so bitter, that they yearn to forget it and obliterate any reference to it. Those who survived the agony of the Holocaust were, for years, numbed into silence. The victims of the Partition of India, on both sides of the divide, found nothing redeeming in remembering and retelling its horrors.

In recent memory, one country after another – Ireland, Rwanda, Argentine, Chile, South Africa – had to face the meaning of vengeance and justice. "Without memory, there is no healing," Archbishop Tutu, the Chairman of the Truth and Reconciliation Commission, observed. "We remember so we can forgive. Without forgiveness, there is no future."

Through this process, for the first time in human history, South Africa said 'No' to the reprisals and counter-reprisals that have been typical of all human conflicts down to our own times, in Rwanda, former Yugoslavia and Afghanistan. "Forgiveness is not nebulous, unpractical and idealistic," Tutu insisted. "It is thoroughly realistic. It is *realpolitik* in the long run."

However, the real theatre of war for most of us is the family. A brother is slighted and he walks away from his brother, not to turn back for years, sometimes never. Estranged cousins and uncles, spouses and lovers, meet each other at funerals in cold silence. Out of our shallow pride we pretend that we don't need each other, and we are self-sufficient unto ourselves.

"What is so special about being a brother?" asks a man in anger. "Just because we came out of the same womb?"

One can find a million reasons for trivializing the human destiny and shattering the fragile bonds that hold us together. But it is only through the courage to embrace one another again, and yet again, that we can hope to discover who we are and what we might become.

Those who exhume the dead bodies, or turn the ashes over in a funeral pyre, may find some evidence of a tormented history. But ultimately one must turn to the living, with all their frailties and one's own, for any solace of the soul. Life itself is a great magical mirror that alone shows that my face, and my brother's face, and the face of my tormentor, are, in the illuminating words of Thomas Hardy, "but one mask of many worn by the Great Face behind."

14

Crushed Flowers and the Smouldering Fire

During the rise and brutal assault of Japanese imperialism in Asia during the 1930s and 1940s, some 200,000 young women from Korea, Philippines, China and Indonesia were used as sex slaves in hundreds of 'Comfort Houses' by Japanese forces. Once the gruesome war was over in 1945, most of those 'Comfort Women' who survived were abandoned by their own families and communities: they were 'defiled' by the enemy; in accordance with the prevailing social values in Asia, what 'respectable' man or community would accept such a 'rotted' fruit? And thus, for decades, it remained a hidden secret, both in Japan and amongst the nations of the victims: Japan denied vehemently that it had ever used any such women; the families and communities of the women endured an unspoken shame that such women still lived among them.

I met four of these women, all in their 80's, in Toronto in 2001; some young bravehearts had decided to tell their woeful tales to the world before they passed away. "Life is an endless recruiting of witnesses," observes Carol Shields in her remarkable novel, *The Stone Diaries*. Quite unwittingly I became a witness to a horror I could have barely imagined: horror of the anguished silence of those whose honour was tarnished because their loved ones were defiled by the enemy.

Through drawings and paintings, interviews and narrations,

sobs and cries, these four victims of the 'comfort' of warriors and the 'discomfort' of their mothers and brothers, rendered a very jagged human face that has so often been lost in the dustbin of history.

But is it really lost? Or does it become new soil for life to give a new birth to itself, to blossom afresh?

Think of hundreds of thousands of women during the war of liberation of Bangladesh in 1971, and how they were brutally assaulted, traumatized and killed. Or the horrors of Rwanda. How much effort and passion go into denying, hiding and trivializing what is so patently and overtly brutal, inhuman and dehumanizing. A painting like *Guernica* by Pablo Picasso, or Edvard Munch's *Scream* may evoke some of that pain, but the deep pond of memories gets murkier and muddier by the day. And one wonders how to wade one's way through it.

In every family – indeed in every community and a nation – there are certain secrets and lies that remain buried, sometimes for decades and centuries. There are times when these skeletons in the closet are never unearthed; they become part of the dust in the grave, or the funeral pyre.

Do such skeletons, or their crumbling dust, have any meaning to anyone's life other than the person who guards them against some prying eyes, or the community that vehemently denies their existence? I wonder if it is the unvarnished truth or the guarded lie that is at the heart of life, and of creation.

"*Art* is a *lie* which makes you realize *truth*," mused Picasso once. "The *artist* must know the manner whereby to convince others of the *truthfulness* of his *lies*." When an artist – like Alice Munroe or Munshi Premchand – sets out to observe that life with an ardent commitment and imagination, our being is illuminated far more radiantly than by all the huff and puff that goes into the vanities and falsehoods of the heroes of our proclaimed history.

I think of two very remarkable films by the English director, Mike Leigh. *Secrets & Lies* (1996) and *Vera Drake* (2004), both largely improvised and both about poor, working class women, with their own secrets, and with their own anguished dilemmas. Watching the films reminded me of the family secret that my own mother harboured, and which, to her dying days, she hardly dared to share with me.

In the social turmoil of 1950s, Mrs. Vera Drake performs abortions secretly for the poor and the forlorn, with utmost discretion and without a penny of profit. Her secret extends even to her own family, to her husband and daughter, to all members of her comforting cocoon. When her 'crime' is revealed and she is led to jail, what falls apart, and what is left to re-assemble, are the splinters of a troubled life, of her own, and of many near and dear ones. In her mundane life of compassion, one may search for grace, hope for forgiveness, question the meaning of life, or bemoan its ironies. But one cannot remain untouched by it.

Vera Drake, one might say, lived a life of lies, so graciously and gently hidden away even from her own family; she was certain that they could not endure the brutalities of the truth, or its flimsy and thin veneer of social pretensions. She was an ordinary, and seemingly an inconsequential person. But she was also heroic, so exalted in her moral grandeur.

The other film, *Secrets & Lies*, is also about a woman, who is discovered at a family reunion to have abandoned a child – unwanted perhaps – many years earlier; it is not just any child, it is a black child, of a black father, in a world where, once, there was no cordial, or even remotely acceptable, meeting of the black and white. Once the secret is out, how does life go on? How does one celebrate the twists and turns of the firmaments of time? As *pas de deux* of lies and truth? As a cosmic dance?

All art, in myriad ways, is meant to lead us to ourselves, to that bed of ashes, where so much lies buried, and where fire still smoulders. It is thus that I come to the secret that my mother harboured, and of which she spoke only once or twice, and in the most hushed of voices. The secret was not about an abortion or an illicit liaison. It was not about her or about any of her children. It was about a distant cousin of mine in India – a pretty girl, as I recall from my youth, a vivacious girl, growing up in abundance and prosperity. As was the custom, her marriage was arranged into an equally prosperous family in Delhi. On the wedding night – and here my mother broke into quiet sobs – the poor girl discovered that she was in the company of not only her husband but a group of his friends, perhaps his clients. She was to be the newest and the freshest member in a ring of call girls in the thriving metropolis.

"But, Mother," I protested in rage, "How can this be? Couldn't she run away? ... What about her parents? Her brothers? The police?"

"Once so defiled, who was going to accept her?" my mother sighed. "They cried and cursed their fate. But they didn't want anyone to know about it..... It was such a disgrace."

In hushed tones, in quiet sighs, my mother was carrying the burden of a thousand years of history, a history riddled with secrets and lies, and yet always pretending to be the truth.

15

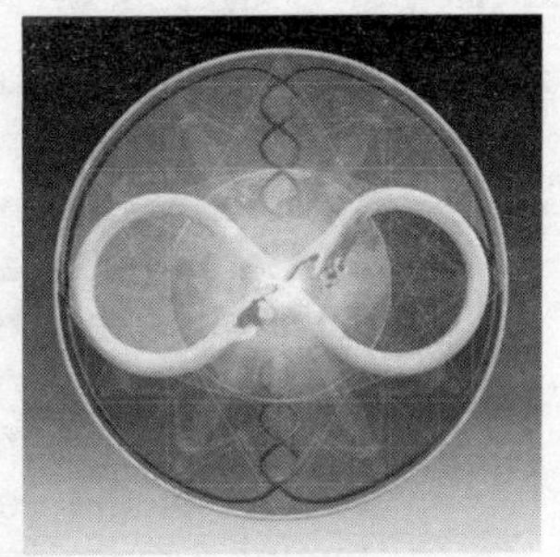

One Face, Many Masks

In 1493, the great genius of Italian Renaissance, Leonardo da Vinci, was commissioned to prepare a fresco in the refectory of the Church of Santa Maria delle Grazie in Milan. Over the next five years Leonardo created one of the great works of art, *The Last Supper*, depicting in exquisite detail the last meal Jesus had with his twelve disciples before being delivered into the hands of the Romans and ultimately led to the Cross. The artist presents that specific moment when Jesus says to his disciples: "One of you will betray me," and the disciples, including Judas, deny in utter vehemence: "Surely, it is not I, My Lord?"

The making of this exquisite fresco, as much of Leonardo's life, is steeped in legend, and in fables and myths that have lasted for more than five hundred years. Whether these fables are lies or half-truths, one cannot say with any certainty but they do reveal a certain kind of truth about human longing for redemption. It is said that Leonardo, known for his chronic procrastination and eccentric temperament, greatly agonized about portraying the central figures of the painting: Jesus and Judas: How to portray these two diametrically opposite cosmic forces – of infinite love on one hand and of betrayal and treachery on another?

In truth, who has ever seen Jesus or Judas except in one's mind's eye, through imagination? So, according to the legend, Leonardo roamed the streets of Milan looking for a man who could serve his model for Jesus, searching for someone who appeared noble and radiant, and loving and compassionate. After much search he did find such a man who was flattered to be asked by the great Maestro to model for Jesus.

And then years rolled by; the fresco was nearing completion except for the depiction of Judas. Where was Judas to be found? A man who had betrayed his Master for thirty pieces of silver. A mean man. A treacherous man. Ugly in his soul. Ugly in his face.

Leonardo roamed the streets of Milan once again looking for such a man. After many days of search he found just the right man – disheveled, drunk and mean, incoherent and shifty. "Would you model for me as Judas for The Last Supper?" Leonardo asked him. "Of course, Maestro, of course...Flattered ... flattered to be asked." And started following the artist to the Church. After a few minutes, the man tugged at the artist's coat and asked for some coins for a drink. As they turned another corner along the narrow lanes, the man tugged at Leonardo yet once again: "Maestro, perhaps you didn't recognize me. Four years ago, you had asked me to model for Jesus."

Jesus. Judas. How very difficult it is, even for a great artist, to see that essential something that is inherent in both of them, in all of universe. In Leonardo's fresco, on Jesus' right is the swooning John (what in Dan Brown's *Da Vinci Code* is presumed to be Magdalene Mary, the mother of Jesus' unborn child), and then, next to him, is the dark-faced, somewhat shrivelled Judas, with his bag of silver in his right hand. Indeed, Judas *is* made to look different from others; how else would *we* know him?

This remains forever our most formidable spiritual challenge: How to see Jesus in Judas, to know them as one? How to recognize that level of being that is beyond good and evil? How to know that no one is beyond forgiveness and redemption? "What compels us to assume," asked Friedrich Nietzsche so disarmingly, "that there exists any essential antithesis between 'true' and 'false'?"

Why does a Judas lurk so closely and so intimately next to Jesus? Is this dilemma eternally planted in the human heart? Are we always to perch on the seesaw of Jesus and Judas, weighing in on this side one moment and on the other the next? One can trivialize this dilemma or celebrate a certain human longing that yearns to see them in some fragile bond together, or even as one. "What is done out of love," insisted Nietzsche, "always takes place beyond good and evil." In that penumbral region, in a certain state of grace, the face of Jesus and the face of Judas, become, in the illuminating words of Thomas Hardy, "but one mask of many worn by the Great Face behind."

How often, in our notion of justice, in the face of the wrong-doer, we see nothing but 'a crime', or even evil? We punish the wrong-doer; his friends abandon him, believing that he is defiled, and any contact with him will prove contagious. Sometimes we seek revenge: 'An Eye for Eye', we proclaim with just pride. Sometimes we pity such a person, and take pleasure in humiliating him. We even vaunt our own virtue, quite certain that the other is irredeemable, or at least incorrigible. Gradually but surely, he is excommunicated, thrown out, exiled from our circle. Rarely do we have the grace and the generosity of heart to recognize the 'Banality of evil' – the spontaneity of a wrong act? And often, perhaps, a little more than the sheer momentary foolishness of it?

'Banality of Evil' is a phrase coined by political theorist Hannah Arendt; she attended and wrote about the trial of the infamous Nazi, Adolf Eichmann, in Jerusalem in 1960s. In her book, she suggested that great evil deeds in history and during the Holocaust specifically, were committed not by fanatic, evil people, but by ordinary people, blinded by the power of the state or an evil ideology. There are many who have argued against Arendt's thesis insisting that there was nothing banal about Eichmann and others who committed such horrific atrocities, and over many years. They were passionately calculated acts of evil.

There is, it must be recognized, great and significant distinction between horrific acts of evil and violence – as in the Holocaust, or in Rwanda, or in Stalin's gulags – and every day acts in the theatre of ordinary life of betrayal, foolishness, indiscretion, trivial transgressions, emotional outbursts, thoughtlessly followed temptations, and a thousand other 'crimes' that are indeed banal. These are the acts that consume us and weigh heavy on our hearts most of the time. To forget and forgive such acts, I believe, would be very nurturing, and would be an act of true grace.

Dr. Israel Charny, an Israeli psychotherapist and Editor of The Encyclopedia of Genocide, once observed: "Everywhere I looked in history, it was the same. The Germans were the essentially the same as all peoples – and ultimately were the Jewish people. None of us are immune from the danger of possibly becoming mass killers, and none of us are entirely lacking in humanity and decency. Whatever it is that makes people monstrous destroyers is potentially in all of us. The seeds of the cure, it seemed to me, were also potentially in all of us."

How rarely we see such seeds!

The celebrated Russian poet Yevgeny Yevtushenko saw these seeds in 1941, at the age of 8, in the faces of war-ravaged women in Moscow as they came out to watch a march of 20,000 German prisoners of war. Each one of these women had lost a brother, or a son, or a husband to the Germans. The women saw the emaciated blood-stained soldiers hobbling past on crutches, or leaning on each other. "Then I saw an elderly woman in broken-down boots push herself forward," the poet wrote in 1962. "She went to the column, took from inside her coat something wrapped in a coloured handkerchief and unfolded it. It was a crust of black bread. She pushed it awkwardly into the pocket of a soldier, so exhausted that he was tottering on his feet. And now, suddenly from every side, women were running towards the soldiers pushing into their hands bread, cigarettes, whatever they had."

It is not known if, as he was working on his great masterpiece, Leonardo's encounter with Judas/Jesus on the streets of Milan – if it was real at all – was, in some ways, reflected in the immortal fresco that he created. However, it is well known about the eccentric genius that he would often purchase caged birds at the local markets and set them free.

As well, when he passed away in 1519 in France, even as princes and great artists mourned Leonardo's death – and created new, and quite untrue, legends about him in magnificent paintings hundreds of years later – in accordance with his will, sixty beggars followed his casket. Such a motley crowd of half famished, ill-clothed and diseased men and women, though undoubtedly spectacular in its own right, has never been the subject of any work of art.

16

The Gift of Fearlessness

The Buddha stands in *abhaya mudra* bestowing the gift of fearlessness; *dukha* – suffering, fear, anxiety – he said, permeates all life.

What anxieties? What fears? Fear of not finding enough food to survive. Or, the fear of being turned into food for someone else. Even a great lion is afraid of going hungry; with all his fierce powers, he cannot command, or cajole, or entreat anyone to bring food to him.

Our life indeed is steeped in a myriad fears; for millennia, hunger and famine, wars and droughts have stalked our destiny. In the Book of Revelation there is a story of the Four Horsemen of the Apocalypse, each horseman personifying pestilence, war, famine and death. In 1964, a remarkable Czech film by Zbynek Brynych, probed the debilitating destruction of the hearts and minds of the Jews in a state of terror and fear in the Nazi-occupied Czechoslovakia; it was named so tellingly ....*and The Fifth Horseman is Fear.*

Fear indeed has an apocalyptic power; like a mighty river that is frozen in winter, fear can sap all energy, and make everything inert and lifeless. But what fear? There is fear of failure. Fear of feeling embarrassed. Fear of being laughed at. Fear of loneliness and abandonment. Fear of ignorance. Fear of mediocrity. Fear of being ostracized. Fear of old age and disease. Fear of death. Fear of feeling unloved. Fear of poverty and hunger. Fear of being lost. Fear of rejection. Fear of being bullied. Fear of humiliation. Fear of being found guilty. Fear of one's dreams going awry. Fear of being nobody. Fear of unworthiness. Fear of being betrayed. Fear of betraying those who trust us. Fear of a life lived without any meaning, without purpose.

Sometimes others' fears seem to us so totally unwarranted, even as we nurse our own fears as necessary and justifiable, even utterly natural. I remember how my mother was petrified by the fear that if in her old age she would survive my father, she would be treated with utmost neglect and humiliation by us, her children and grandchildren. She had good reasons for her fears, and I often thought how not to let this happen – not that she be treated better if she survives my father, but rather that she does not survive my father. Before this crippling fear her other fears seemed quite ordinary, even somewhat amusing. In the presence of certain 'society women' in the new city where we had moved to, she made every effort to hide that she did not know how to read or write at all, or worse, that she could not even sign her own name. Or even, sometimes, that she did not understand or speak a word of English. Her husband, my father, was an esteemed lawyer – a *vakil;* and so, she was to be called *vakilini*; at a time when there were no women lawyers, vakilini could mean only the 'wife of a lawyer' and not a female lawyer. And so these various fears were expressed or hidden by her, sometimes clumsily, and sometimes cleverly. Today I can say how very human my mother's fears were, but in my youth they seemed hypocritical, even cowardly.

In some circles the panoply of mortifying fears has been called 'existential angst', as though life itself is a source of fear. How can one live with full intensity and with all ebbs and flows without knowing what fears freeze the mighty river of one's life? It is thus that one yearns for the gift of fearlessness, for otherwise one can go all through one's life carrying the burden of one fear or another, shrunk in one's being, living in a 'winter of discontent', unfulfilled, entrapped in *ifs* and *buts*, without flow, without harmony.

A few years ago, on a hot, sultry weekend, some 150 men and women had gathered in Toronto to learn about some of these shackles of fear, and how how they might be broken. I was among them. As hours rolled by, a frail and nervous woman stood up to share her story of fears with us. She was 53, and a grandmother. "I have never dared to share this before ... I feel dirty and burdened. I was four when it happened...My grandfather abused me sexually..." she sobbed as she poured out a seething stream of hot and broken words: "It went on for many years...he died over 20 years ago. But I see him all the time ...in every man, even in my son. Until now I have never spoken a word about it for fear of humiliation... I have never forgiven him. I have never forgiven myself for keeping silent..."

It was a hot day, but a cold pain seized many hearts, obscuring any sense of life's warmth. For this woman, this dreadful memory and the fear of being found out had inevitably travelled through the convoluted corridors of her mind for 50 years, and they would surely accompany her to her grave some 20, 30, 50 years hence unless she learnt to break their spell and their power.

A hundred years ago, with the rise of Freudian psychology, one began to believe that many fears and phobias have their origins in early childhood. In order to address them, and to subdue them, with the assistance of a therapist, through memory recall, one must travel back to those early years. In the last decades of the twentieth century, some journeys back into the farther reaches of one's subconscious, however, led to awful fabrications of memory, where the distinction between what was 'real' and what was only made to imagine, often by zealous therapists, out of anger or humiliation or sheer ideology, was completely obliterated. In this false memory recall – turned into an ugly hysteria that led to the ruination of lives of thousands of parents and citizens – there was no place for forgiveness, or for divine grace, or for last judgement, or for acceptance of human folly. It was as though, in a hall of mirrors, memory played monstrous games that turned anxiety into fear and paranoia, and then into retribution against imaginary demons.

How do we know then what fears are real, or when they are a mere phantom of the mind, such as fear of flying, or fear of strangers, or fear of water? Fear comes from somewhere inside, from memory, from an unpleasant experience, from a thwarted hope. It must come from the fact of life itself, from the being and becoming of life.

It would be very satisfying to believe that the genetic alphabet writes the story of our lives and mixes a pinch of fear in it. I wonder if we know what life is, where and when and how it all starts, and where it all goes. The existential angst may be wrapped up in these and in many more such uncertainties. Many of us, for millennia, believed in *karma* and *samskara* – the ideas that a being is in the making over many lives, and one's thoughts, actions, intentions, yearnings and exasperations are all part of an on-going formation, reformation and transformation. In this drama there may be genes and memes, defining some of our biological and cultural inheritance, but each character in the drama remains uniquely its own. It is this special unexpected uniqueness that keeps our universe forever indefinable, evolving and creative. A therapist may, if one is fortunate, throw a little light on this riddle, but the ultimate healer must be each man himself; he alone must conjure up the powers that lie coiled in him somewhere.

I was once attending a show in Disney World with my 8 and 10 year old daughters, Ankita and Amrita. It was called 'Honey, I Shrunk the Audience'. Wearing special goggles, we experienced the fear of tiny, menacing mice running at our legs, vicious vipers lurking about us, and overgrown dogs ready to devour us. It was so very frightening and thrilling, and all so very *real.* All of us – children and adults – were screaming with terror and excitement, except the 8-year old Ankita. Once the show was over, and we had goose bumps as witness to our experience of fear, I asked her, "How come you were so calm; you were not screaming at all?" She said calmly: "Dadu, it was quite simple. I took my goggles off."

Today I reflect on it and wonder what goggles we have put on, sometimes unknowingly, sometimes wantonly, and sometimes capriciously, that continue to create for us this extraordinary drama of fear and morbidity. And how we must keep these goggles off our eyes in order to shut out the deadening fear that seems so grippingly and unmistakably real? With what hands – and what eyes – do we receive the gift of fearlessness of which the Buddha spoke more than 2500 years ago, much before Sigmund Freud or Jean-Paul Sartre?

What therapist do we seek who can show us what is real and what is merely a phantom?

17

When Death Plays Chess

For millennia, brave young men have sacrificed their youth and their valour in all kinds of wars, inspired by imperialism, by religious or patriotic frenzy, by fear of an unseen enemy, by dreams of triumph over the heathens and the *kafirs,* or by sheer barbarism. So it was during the long years of the Crusades that engaged the Muslims and the Christians in fierce battles over many centuries in the Holy Land.

One Christian crusader, as he marched to Jerusalem to earn religious indulgences, and to liberate it from the Muslim infidels, wrote home: "[We] made an attack upon them and killed an innumerable multitude. All the others fled in confusion. Our men, moreover, returning in victory and bearing many heads fixed upon pikes and spears, furnished a joyful spectacle for the people of God."

A joyful spectacle it certainly must have been!

The great classical Swedish film, *The Seventh Seal*, is the story of one such man returning home from the Crusade, not with 'heads fixed upon pikes and spears,' or with any 'joyful spectacle for the people of God,' but with a devastating, corroding sense of futility and emptiness. For ten years, this knight did what he was called upon to do – kill, maim, burn, rape – all in the name of God, all with the hope of finding some solace in that calling. And now returning home, to a land devastated by plague, he wonders where this God is who had so commanded him, and why he remains so hidden, so obscure.

When one raises such a question, or worse, is gripped by such an enquiry, one is staring into an existential abyss; one wants to know why one has taken the journey, where one's steps went astray, and where they might end.

> Knight: *I want to talk to you as openly as I can, but my heart is empty.*
> *... The emptiness is a mirror turned towards my own face. I see myself in it, and I am filled with fear and disgust.*
> *.... Through my indifference to my fellow men, I have isolated myself from their company. Now I live in a world of phantoms. I am imprisoned in my dreams and fantasies.*
> *.... Why can't I kill God within me? Why does he live on in this painful and humiliating way even though I curse Him and want to tear Him out of my heart? Why, in spite of everything, is He a baffling reality that I can't shake off?*
> *... My life has been a futile pursuit, a wandering, a great deal of talk without meaning.*

A life without meaning, yet a tenacious, obstinate desire to hold on to every breath. The plague-infested land, to which the Knight has returned, is reeking with death. Death is lurking all around him, yet to seek reprieve from it, however momentary, the Knight challenges it to a game of chess.

A game of chess!

No other game is marked by such ingenuity and cunningness as chess; every move opens many more; at every step there are many pathways, and soon the spreading delta of possibilities seems infinite. One can see only so far ahead, only in so many directions and ways, one can recall only that much. Soon it becomes a dense forest, without a ray of light piercing through the thick canopy. A game of chess with death then is doomed to defeat. A reprieve may be, but a victory, no.

How does one stare death – futility, absurdity, devastation – in the face and find some consolation in the living?

In 1954, when the memory of horrors of India's Partition were still fresh in the minds of many, a young man of 28, Dharamvir Bharati, wrote a play, *Andha Yug* ['Blind Age'], which has come to be regarded as one of the great plays in Indian theatre. The play is set on the last day of the fierce battle in the great epic of the *Mahabharata*. The battle between the two families has raged on for many a long day; in its wake everything and everyone is destroyed; worse, everyone has succumbed into cynicism of a blind, dispirited age, that has so desperately lost all its moral moorings. No one is certain anymore what is right and what is wrong. All one can see, as far as the eye goes, is death and the vultures in the dark sky.

Our human story has been replete with such cynicism and moral decay again and again; on August 5, 1994, when this play was performed at Tonga in Japan, it commemorated yet another act of vicious blindness that unleashed the atom bombs at Hiroshima and Nagasaki. The play, however, like all acts of courage, refuses to abandon Hope:

I suddenly understood
as if in a flash of revelation
that when a man
surrenders his selfhood
and challenges history
he can change the course
of the stars.

The lines of fate
are not carved in stone.
They can be drawn and redrawn
at every moment of time
by the will of man.

The father of Atom-bomb, Robert J. Oppenheimer, proved to be another Knight in the *Seventh Seal*. When he witnessed the tests of the bomb in July 1945 in New Mexico, he recalled a line from the Bhagvada Gita: "I am become Death, the shatterer of worlds."

In the epic the words are of Sri Krishna, the Exalted One,

Lord of the fate of the mortals, the charioteer of the great warrior, Arjuna. On the tongue of humans, however, the words become menacingly foreboding. Man was now in possession of a mighty new instrument of darkness.

After the bombs were exploded, and the horrific tales of destruction began to pour in from Hiroshima and Nagasaki, a doubt began to take root in Oppenheimer's conscience. "Have I created a monster?" he wondered.

"What have I done?" he bemoans. "How can I set it right?" He repents; he begs for forgiveness. He suffers at the hands of authorities. He is thrown out of a powerful community of fellow scientists, many of whom are now developing even more powerful weapons of destruction, and can brook no doubters of relentless progress of science. Oppenheimer is crushed by the Military Industrial Complex that survives and thrives, as a grand merchant of death, by manufacturing and selling more and more insidious weapons of destruction.

Oppenheimer's dilemma played itself out, I believe, on a much wider canvas than did Arjuna's in the Mahabharata: How to engage the creative prowess of one's faculties, for life or for death? How not to be consumed, not only by a hot and seething war, but also by a cold war of nefarious ideologies and insidious pretensions? Oppenheimer was faced with the extinction of the whole family of man, even possibly all of life; he was faced with an enemy that lay within his own camps, in his own laboratories, and in his own home, and thrived on being the seductive merchant of death and darkness.

Again and again, death masquerades in many garbs, it plays many games. Again and again it takes us to the brink of an uneasy, tormented life and wraps itself in many enigmas. In its eternal game of chess, it asks yet once again: "So, what's your next move?"

On the morning of August 6, 1945, a young Japanese engineer, Tsutomu Yamaguchi, was just stepping off a train in Hiroshima on his business trip for Mitsubishi Heavy Industries, when the city was hit by the first ever A-Bomb; its ground zero was three kilometers away but Yamaguchi, like hundreds of thousands others, were devastated by the after-effects and the radiations that shot forth. His ear drums were destroyed; he was blinded temporarily, and was left with serious burns on the left side of his body. He was wrapped in bandages for his skin wounds, and he spent a fitful night in an air raid shelter before returning to his hometown 260 kms away.

Next day, on August 9, at 11:02 am, as his boss reportedly questioned his sanity for claiming that a single bomb had destroyed a whole city of the size of Hiroshima, another A-Bomb exploded above him in Nagasaki, throwing Yamaguchi to the ground. He had walked in the path of death twice, and had mysteriously come out alive both times. At the age of 93, Yamaguchi is believed to be the only man to have been near Ground Zero both in Hiroshima and Nagasaki, and to have survived.

Such strange and unexpected things happen, and we don't know how or why. We wonder if life is not an inscrutable game of chess.

We wonder if we are the the players or the pawns.

❖

18

Why God Won't Go Away

If God were a real person, or something concrete and tangible, and 'natural', with all the vociferous attacks and assaults on it in the past several years, it could be mortally wounded. For many God is indeed a person; he speaks, blesses, walks, watches, punishes, loves; a certain book is his word; a certain song has been sung by him. For some he may be eternal, transcendent and immortal, but still he was born on a certain day, and he has a place of birth. He may be a mighty force in the universe, but for many, he has a face; he has eyes and ears. He is a 'person'; he is 'natural'.

Many years ago, in 1927, Bertrand Russell wrote his famous book, 'Why I am not a Christian', and Mahatma Gandhi, in turn, wrote, 'Why I am a Hindu', each talking of his religiosity, or his aversion to it, in the framework of his own thoughts and experiences, and how God was to have no place in his life, or was so profoundly essential to it that nothing else mattered.

Russell spoke for millions all over the world who abhorred all that went on in the name of religion:

> Religion is based, I think, primarily and mainly upon fear. It is partly the terror of the unknown and partly, as I have said, the wish to feel that you have a kind of elder brother who will stand by you in all your troubles and disputes.... A good world needs knowledge, kindliness, and courage; it does not need a regretful hankering after the past or a fettering of the free intelligence by the words uttered long ago by ignorant men.

Gandhi, on the other hand, drew his inspiration for a non-violent struggle for the independence of India, and of a new vision of India, from Hinduism:

> I am a Hindu because it is Hinduism which makes the world worth living. I am a Hindu hence I Love not only human beings, but all living beings.

Yet, in his own inimitable way, when asked if he was a Hindu, Gandhi also said:

> Yes I am, I am also a Muslim, a Christian, a Buddhist, and a Jew.

Of late, many strident and eloquent voices have declared God to be no more than a delusion; in contrast to the chants of *Allah-hu-Akbar*, they have insisted that he is not great, and his spell must be broken. This 'New Atheism', as it has come to be known, mocks the claims in various scriptures about the creation of the universe, or the divine origins of man; it derides some of the biblical writings or the passages in the Koran. Biologist Richard Dawkins has dubbed the teaching of religion to children as 'child abuse'. Noble Laureate physicist Steven Weinberg warns that "the world needs to wake up from its long nightmare of religious belief. Anything that we scientists can do to weaken the hold of religion should be done and may in the end be our greatest contribution to civilization."

As a scientist myself, and as a student of the history of science and cultures, I am sometimes called upon to take sides, or at least comment upon this raging battle. I envy the sheer steadfastness, bordering on evangelical fundamentalism, of people like Richard Dawkins, who insist that the only true path to human salvation is through science. "Darwin provides a solution, the only feasible one so far suggested, to the deep problem of our existence," he proclaims. "We no longer have to resort to superstition when faced with deep problems: Is there a meaning to life? What are we for? What is man?"

In the same vein, biophysicist and co-discoverer of the structure of DNA molecule, Noble Laureate Francis Crick presents an 'astonishing hypothesis' in his book *The*

Astonishing Hypothesis: The Scientific Search for the Soul: "'You', your joys and your sorrows, your memories and your ambitions, your sense of personal identity and free will, are in fact no more than the behaviour of a vast assembly of nerve cells and their associated molecules."

This is indeed an astonishing hypothesis; as far as I know there is no scientific proof for it, or any other kind of proof. It strikes me as unreasonable and most profoundly fatalistic. Any notion of "no more than" gnaws at me.

Atoms and molecules may be the alphabet of the universe, but are Shakespeare's soliloquies a mere ensemble of twenty-six letters of the English alphabet? Does the eloquence of poetry come from the alphabet or from a special arrangement of words, and their allusions and mythopoeic resonances? Is the great Chartres Cathedral nothing but stones? What makes it a work of such grandeur in contrast to another building made of similar stones? Does a film on the screen that makes one moan with pain because of one's empathy for a character nothing but flickering of light?

This belittling of the universe and its magnificent creative grandeur and of life and its myriad experiences is to understand science in a most reductionist manner. In the true spirit of science, I should say, that just because these assertions are made by the learned and the celebrated scientists, they should not discourage us to questioning their vehemence.

At no place did Darwin talk about the 'meaning of life', though he wrote extensively and profoundly about the origin and evolution of life. He wrote instead about ethics that must guide human conduct, and how natural selection seemed to no longer act upon civilized communities in the way it did upon other animals. He called sympathy "the noblest part of our nature", and that it must not be checked "even at the urging of hard reason." Forces of nature – gravitation, electro-magnetism – have no human face; they do not act morally or ethically, as we humans understand these words. They are simply 'natural'. Human urgings, on the other hand, are often trans-natural; they leap into a universe that is touched by a whiff of the miraculous, as though some fibres of the unexpected and the magical are woven into it, however slightly, however ephemerally. That is how for some the universe is a great dance: "O Body, swayed to music, O quickening glance/How shall I tell the dancer from the dance?"

As a counterpoint to the attacks on God and religion, of late there have been other books as well, some written by neuroscientists, who claim that God has some neurological roots; God is not in our hearts and souls but in our brains. One book, provokingly asks in its title: 'Why God Won't Go Away: Is the New Atheism Running on Empty?' Another is titled: '*Born to Believe: God, Science, and the Origin of ordinary and extraordinary beliefs*'. Still another is titled: *'The God Gene: How Faith Is Hardwired into Our Genes*'. "Many theories try to explain," say these books, "the psychological and sociological reasons why people nurture spiritual beliefs, but the answer is found in neuroscience – indeed, in the very synapses of our brain."

Like an earlier assertion – "Violence is in our genes" – there seems a very strong tendency at present to place all phenomenological expressions of life, in all their abundant variety, in the network of neurons. By taking pictures of brain activity in our various states of being, neuroscientists wish to prove irrefutably and scientifically that God or hatred or love or heedlessness or indifference are all in our neurons. "Here, you can see it," they declare triumphantly.

I have a disquieting feeling that such claims, and some deductions made from them, don't always make very good science. A particular state of our mind – of anger, or frustration, or peacefulness – may be reflected in our blood pressure and in other measurables. However, the latter do not tell us why we get angry, or don't always seek peacefulness. What *God Gene,* I wonder, can guide us to seek consolation of God in our moments of suffering? What kind of God would that be?

God may be in danger in some places, but it has had a long life in most places of the world: Whether in Ireland or in India, in Mexico or in China, some notion of God, sometimes wrathful, sometimes consoling and compassionate, has thrived for millennia. With the coming of science – the ushering of the Age of Enlightenment, of reason and rational thought, and the discovery of natural laws – many had hoped that God, and its million attendant institutions, rites and priests, would die a natural death. But it still lingers on quite tenaciously, and can be said at times to be thriving.

To be sure, now we rarely seek intervention of priests for the cure of malaria or schizophrenia, or waft some blessed smoke to serve as a way of avoiding an automobile accident; we demand healthcare for all, and better machines. Fewer and fewer people in the world now believe that the life of inordinate wealth and privilege of some has been ordained by God; or equally, that poverty and misery of millions is part of some divine scheme. In his 'Wealth of Nations', Adam Smith, or Karl Marx in the *Das Kapital* searched for material social and economic forces to account for the miseries of the world. In their new vision of social morality, God had no place; he was in fact an obstruction, and religion was 'opium for the masses'. Certainly over the decades, in many places, the dominion of the kingdom of God, and of his minions, has shrunk a lot.

Yet God – whatever that inscrutable and infinitely pliable word may mean – will, I believe, live on, not because it is in our genes, and we are 'wired' for it in our neurons, but because of life's profound uncertainty and unexpectedness, and because of our longing for meaning and coherence. No amount of science and no theory in physics can predict the unfolding of life even for a moment. What thoughts will whirl across our mind at any instant, what events and images will present themselves to anyone of us at the next moment, still remains uncertain and full of new potentialities. Do we not live in an unexpected universe, which occasionally lends itself to our desires and dreams, but which, time and again, presents itself as new and quite miraculous?

As well, no amount of science renders any meaning to life; all science asserts again and again is that that there is no meaning, and there is no purpose to existence.

In the state of uncertainty and existential void we sometimes seek guidance, we seek shelter, we seek consolation, and we seek love. We seek Hope in this universe of infinite possibilities. Anyone who holds our hand when we are lost is a messenger of God. Anyone who holds us in an embrace when we are lonely takes on the body of God. Anything that shines a light in our eyes when we are losing sight is a Star of David. It is thus that God manifests itself. For some it may be a delusion. For others it may be the only reality.

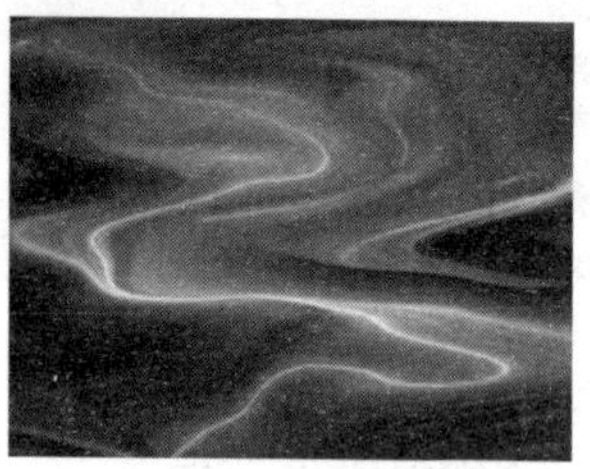

19

The River of Life in the Valley of Death

Death knocks at many doors; often it is at a stranger's door. We hear its knock with pity. We feel sympathy or horror, or even indifference. People die but life goes on. It is as though the mighty river of life knows no end; it flows on and on, unceasingly.

But when death knocks at our own door, its sound is different. The sobs and sighs it evokes are different. The pain it triggers is different. The void it creates is different.

That is how it was for me when my brother passed away suddenly. In these two words, *passed away*, a whole world closed on itself; a snowflake melted; a light was extinguished. A deep, foreboding silence had descended. What was a living, thriving presence a day earlier had now turned into a heap of ashes. To the incantations of "Soul is immortal", "*Ram Naam Satya Hai*", we submerged the ashes in the holy Ganga at Haridwar. The great, mighty river took it all in, as it has done for millennia, and continued to dance her way to the ocean, a thousand miles away.

Even though in this past decade or two, for many of us the world has expanded beyond measure, the drama of our lives still plays out most intensely and most vividly in our families. It is there, through our siblings and children, our uncles and aunts, our parents, our in-laws, that our world defines itself, and finds its contours and its shores. It is in these shores that we find a harbour, a refuge from the turbulences of the *bhava sagar*, the ocean of our being.

We may call it a network or a tribe, a clan or a family. Whatever name we may call it by, and whoever we are, we need *it* – this web of unconditional ties and knots – as desperately today as perhaps we did ten or a hundred thousand years ago, when we wandered the earth in search of food and shelter. Though sometimes uncertain of our roots and of our moorings, we did nevertheless have some primordial sense that the blood that flowed in my veins was somehow linked, in some special way, to the blood that flowed in the veins of my brother.

"Family life is a fact that underlies everything else about man," writes evolutionist Loren Eiseley. "His capacity for absorbing culture, his ability to learn everything, in short, that enables us to call him human. He is born of love and he exists by reason of a love more continuous than in any other form of life."

"I am my brother's keeper" must have resounded in the very womb of creation, even as, over the millennia, we have struggled to see the face of our brother in the face of a stranger as well.

At Haridwar, one comes face to face with India's great prowess at genealogical record-keeping. Going back to some 200-250 years, over ten generations, we saw the record of our forefathers, and their forefathers, with all their fecund progeny, kept in meticulous detail. The paper has become crisper; the ink has faded; yet *jati* and *gotra* – so integral to Hindu caste system – are still recorded as indelibly as one's blood type today. Long before any birth or death certificates, these records kept by the *kul purohit* – the priest of the clan – were the only, and the most reliable pieces of information about birth and death. Here on the banks of Ganga, as the mighty river flowed on, each human life seems to have left a footprint.

And so it was for my brother; as we symbolically floated his boat on the river, he too now became part of the ever-flowing river of life.

Like many other modern families, our family too is now spread in many countries, and in many cities. And inevitably, like most families, in the midst of our affections, we have seen distances grow between us. We have celebrated our victories and achievements together, but we have not always nourished our tree of life with all the care that it needed. It is the shattering power of 'passing away' perhaps that something that seemed frozen for a while begins to thaw, a new light begins to shine through.

The doors we close and the doors we open each day decide for us the winds that flow through our lives. A life is often many moments of missed glory; a bond could have been deepened; an embrace could have been much more heart-warming. One could have asked for forgiveness, one could have forgotten a slight.

Each moment in our lives is a final performance; there are no rehearsals. There are no reset buttons; there are no erasures. Words are said that should not have been said; there are moments of inattention, of indifference, of thoughtlessness. There are thwarted dreams in every bosom; there is an unrequited love in every heart. "If only," we say, "if only." But there is no going back. The river of life flows only one way.

Before the still smoldering pyre of my brother, I stood distracted by these stray thoughts; in that heap of ashes lay a man that would be me one day; it would be my child on another day.

Is this the end of all our journeys? I wondered. Is this what we fret and struggle for all our lives? Where in these ashes is one to find life's glory? Its meaning? Its purpose? Where?

Time and again, as the river of life meanders in the valley of death, such thoughts are stirred:

Life is real! Life is earnest!
And the grave is not its goal;
Dust thou art, to dust returnest,
Was not spoken of the soul.

.....

Art is long, and Time is fleeting,
And our hearts, though stout and brave,

Still, like muffled drums, are beating
Funeral marches to the grave.

Every day new life is born; everyday life is taken away. But when that life is taken away from my own home, it shows me the real meaning of the river of life. It reveals to me how precious life is; how every moment in every life is precious; how love is what renders flow and vitality to the river of life. And how every man is my brother, how the sun is my brother, and the trees and the mountains too.

So long as life manifests itself anywhere, in any form, death shall have no dominion. Life shall triumph, for it is the most profound and most exquisite expression of love.

That's how I choose to remember my brother in the smouldering heap of ashes, and in the floating boat of flowers in the Ganga!

❖

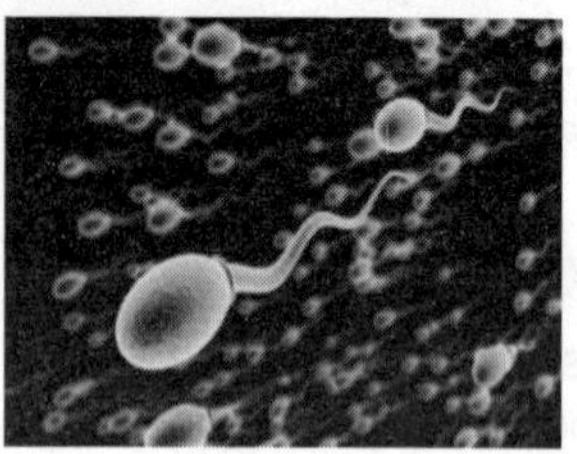

20

Stories Our Fathers Tell Us

There is a story in the great epic *The Mahabharata* about a young archer Eklavya whose skills are found by his guru Dronocharya to be even greater than those of Arjuna. To make certain that no one outdoes his favourite pupil Arjuna, the guru asks for the thumb of Eklavya as his *dakshana*, a sacrificial offering to the guru.

With some minor variations this story has been told all over India in plays, sculptures, songs, dances and paintings for a thousand years; it is told even today as fervently as ever before. It could be said that it is one of the most endearing and enduring, legends of the living traditions of the *Mahabharata*.

It is a moving story, perhaps even a salient feature of the *guru-shishya parampara* of the Indian culture, indicative of the devotion of a disciple to his guru. What is asked of Eklavya is his unconditional devotion; what is offered by him unflinchingly is his most precious gift. There are no moral dilemmas here, but only an assertion of a sacred duty, of supreme sacrifice. Alfred Tennyson, in his famous poem, 'The Charge of the Light Brigade' about a suicidal attack during a battle in the Crimean War (1854-56), captured this supreme sacrificial ethos that has prevailed in many societies:

"Forward, the Light Brigade!"
Was there a man dismay'd?
Not tho' the soldier knew
Someone had blunder'd:

Theirs not to make reply,
Theirs not to reason why,
Theirs but to do and die:
Into the valley of Death
Rode the six hundred.
....
Cannon to right of them,
Cannon to left of them,
Cannon behind them
Volley'd and thunder'd;
Storm'd at with shot and shell,
While horse and hero fell,
....
When can their glory fade?
O the wild charge they made!
All the world wondered.
Honour the charge they made,
Honour the Light Brigade,
Noble six hundred.

Over the decades, Tennyson's words have tended to take on a certain invocational élan, which, in popular parlance, has come to be expressed as: "Ours is not to question why, ours is but to do and die."

All persons in authority, in families and schools, and in all institutions – political, religious, ideological – have commanded loyalty and obedience of their charges. I believe time has come to question such blind obedience and re-vision traditions, whatever their origin, to evoke a different ethos, and to celebrate a different flow of life. It is thus that one must ask: Did guru Dronacharya not violate his dharma in making such an unsavoury demand on his young pupil? Did the pupil, in turn, not show misguided devotion to his guru?

On December 2, 1942, the imperial forces of Japan made a surprise attack on Pearl Harbor; President Roosevelt called it 'the Day of Infamy'; it unleashed the mighty fury of a powerful country on Japan, resulting in the dropping of Atomic bombs at Hiroshima and Nagasaki, and leading to the surrender of Japan on August 15, 1945.

The surprise attack was conducted by young Kamikaze warriors; theirs was a suicide mission. They were sent to kill and destroy the enemy, with no hope of ever coming back alive. These warriors were totally and unflinchingly devoted to their Emperor who was, for them, Divine – god-like, infallible, and above all utterly unknowable. No other man in the world at that time wielded such absolute loyalty and devotion.

As the Kamikaze warriors were being selected for the Pearl Harbor mission, a young man, barely 25, is said to have been rejected because he was married and had a child. Dejected, the young pilot wrote to his wife why he was not chosen for the daring suicide mission. It is said that, in order to remove any obstacle for her husband to make the great sacrifice, she first killed her child and then herself. Her husband was now free of any worldly attachments to make the supreme sacrifice.

If Japan had won the war, in all likelihood, such stories would have been the stuff of folklore, and told to succeeding generations of children in schools and in history books with glorious pride. Today such stories, rightly, suggest little more than blind obedience, to be witnessed, under one hateful name or the other, amongst suicide bombers in Sri Lanka, Palestine, Pakistan and Iraq with the same blind fury.

Do they have the shades of Eklavya in them?

In 1961, psychologist Stanley Milgram conducted a study at Yale University focusing on the conflict between obedience to authority and personal conscience. In his experiments, ordinary human subjects acting as 'teachers', under the command of scientists in white lab coats, continued to administer very severe, though unknown to them, fake electric shocks to their 'students', out of their sight, in another room, until lethal levels were reached, even as the latter begged for mercy. Milgram's experiments revealed a certain relentless power of authority in which individual conscience falls easily by the wayside. Is this how, mass killers like Adolph Eichmann find justification for their atrocities, by arguing that they were only 'obeying orders'? The question that Dr. Milgram posed was: "Could it be that Eichmann, and his million accomplices in the Holocaust, were just following orders? Could we call them all accomplices?"

Experiments like Milgram's, and the questions they raise make us wonder how obedience and loyalty become certain cultural traits that are kept alive and transmitted over the centuries across thousands of miles.

Our stories and myths have travelled a long way in our collective memory; we have been nurtured by them as by our mother's milk. But to replenish a culture, or a tradition, and all the undercurrents that lie hidden in its cocoon, we need to revisit it and re-vision it, so that it does not get fossilized and become merely a moribund habit.

It is through memory that a culture, like a living organism, becomes what it is. A hand has a memory. An eye has a memory. An ear, the tongue, the skin, the feet have memory of hundreds of millions of years of evolution built into them; that is how a human child at birth is destined to stand up at about one year of age and walk, and speak, and create wonders of thought and imagination through words. Communities and nations too have their memories. If those memories are thwarted, or distorted, or are locked up in some inaccessible dungeons of mind, we live shallow lives. Our art and literature become superficial. Our dreams then seem flimsy; they touch nothing, because they come from nothing.

"I grew up in this town," wrote the great poet Pablo Neruda. "My poetry was born between the hill and the river, it took its voice from the rain, and like the timber, it steeped itself in the forests."

All our rites and rituals, our education and our institutions are forever trying to make us obey, to stifle our imagination, and to frustrate our flight to the distant star. With all our frailties, what we need to aspire most is the courage to be. Our being – our soul, our spirit – beckon us to have courage, to have the daring to look life in the face, to probe and explore beyond traditions and ideologies. Our real heroes are those who dared not to follow the beaten path. During his long years in prison, what inspired Nelson Mandela most were the words of freedom written by William Earnest Henley, a Victorian poet who never abandoned his tenacious will to live despite numerous adversities:

Out of the night that covers me,
Black as the pit from pole to pole,
I thank whatever gods may be
For my unconquerable soul.

In the fell clutch of circumstance
I have not winced nor cried aloud.
Under the bludgeonings of chance
My head is bloody, but unbowed.

Beyond this place of wrath and tears
Looms but the Horror of the shade,
And yet the menace of the years
Finds and shall find me unafraid.

It matters not how strait the gate,
How charged with punishments the scroll,
I am the master of my fate:
I am the captain of my soul.

At each step in the human journey traditions have been created by ideologues and idolaters to bind our imagination into orthodoxies. There are no fixed iconographies for the images of Krishna, or of Arjuna or Dronocharya, or of Jesus or Buddha. Or of the brave and the beautiful. These images are forever fluid; they are created in the smithy of our own soul, at every step of our own journey. They are indeed allegorical.

❖

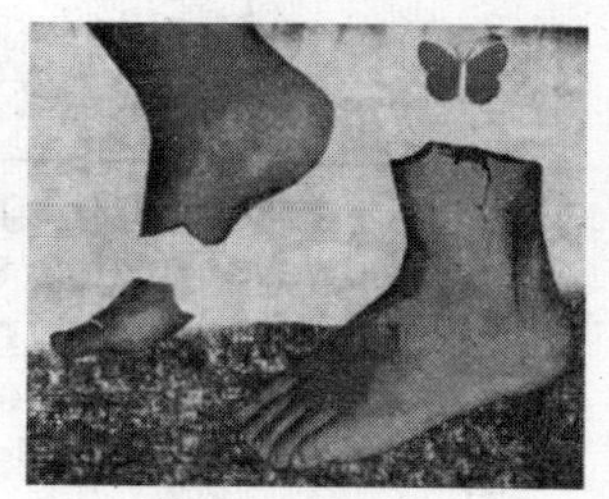

21

Feet of Clay: When Idols Come Tumbling Down

Thomas Jefferson is one of the great figures in American history; his eloquence and liberal republican spirit are evident in the Declaration of Independence: "We hold these truths to be self-evident, that all men are born equal, that they are endowed by their Creator with certain unalienable Rights, that among these are Life, Liberty and the pursuit of Happiness."

One of the Founding Fathers of America, Jefferson is hailed as the finest embodiment of the Age of Enlightenment; he is a man of reason, liberalism and equanimity. However, it is the spirit of our age, of post-modernism, that no man or woman remains unquestioned; greatness, at best, is only partial; any claims to eternal greatness seem far too self-righteous. In case of Jefferson, it is now shouted from rooftops that though he called slavery an 'abominable crime' and a 'moral depravity', he owned hundreds of slaves throughout his life, and he held strong views on the racial inferiority of Africans.

As well, it is now subject of several books and films that for over thirty years, all through his presidency, Jefferson – a young widower at the age of 39 – had an intimate relationship with his slave Sally Hemings, and had six children by her, none of whom ever bore his name.

One of the developments of post-modern age has been

that again and again, 'human nature' and all human beings – however mighty, and however esteemed – are found to have feet of clay. They may be the greatest of scientists or musicians, philosophers of eminence or awe-inspiring statesmen, great educators or artists – Gandhi or Einstein, Picasso or Martin Luther King – but everyone is all too human after all; they have frailties, dark corners, hidden secrets. In other words, no one is without some blemish, without some flaw in his character. It is as though, in the words of Sir Francis Bacon, "There is no excellent beauty that hath not some strangeness in the proportion."

I wonder if it is only the Western liberal ethos that derives some cynical pleasure in exposing the nakedness of the emperor.

Take Isaac Newton, for instance; he strides across the seventeenth century as a god-like figure of extraordinary insights into the functioning of nature. Indeed, English poet Alexander Pope immortalized his accomplishments in the famous epitaph:

Nature and nature's laws lay hid in night;
God said "Let Newton be" and all was light.

Yet Newton is now exposed to have been steeped in pettiness and jealousies; as President of Royal Society, he fabricated letters and documents against his opponents. As Master of the Mint, he sent many counterfeiters to the gallows. As a religious zealot, he wrote more on Biblical hermeneutics and occult studies than he did on science and mathematics; his alchemical experiments may even have led to his death by mercury poisoning. Above all, he was dour, sullen and profoundly paranoid.

In his own judgment though Newton was an eternal seeker of truth: "I do not know," he wrote in a later memoir, "what I may appear to the world, but to myself I seem to have been only like a boy playing on the sea-shore, and diverting myself in now and then finding a smoother pebble or a prettier shell than ordinary, whilst the great ocean of truth lay all undiscovered before me."

In 1970, a very remarkable Italian film, *Investigation of a Citizen above Suspicion*, presented an exposé of political corruption – all too familiar to people all over the world – but with a twist: here a police inspector, after having murdered his mistress, plants numerous incriminating clues everywhere that could lead any investigating officer to the inspector himself. But either out of fear of authority, or out of immense trust in the infallibility of authority, even in the face of the most obvious evidence, every probing eye turned and looked somewhere else. The inspector was a 'citizen above suspicion', or one might say, above law.

Such has been the power of trust in those who wield power that voices of protest are not even raised, or if they are raised, they are hushed at every turn. They are ignored; they have no credibility; they fade away unceremoniously.

In 2010, it almost happened in Canada. Col. Russell Williams was a much decorated and admired military commander of Canada's largest air force base at Trenton, Ontario. For all intents and purposes, he was confident that he was a 'citizen above suspicion'. Underneath his calm and confident exterior lurked a dark and shadowy world of rape, violence and humiliations. He frolicked like a ghost between the two worlds, apparently confident that the two would remain forever apart and unknown to each other.

The principle of legal egalitarianism – that everyone is equal before the law – has made it possible to reveal, indeed expose, the seedy side of lives of those who had the wherewithal and power to hide their acts from the eyes of the law, and of everyone else. Whether it is President Bill Clinton or Emperor Leopold of Belgium who unleashed terror in Africa, or the former President of Israel, Moshe Katsav, or the Managing Director of International Monetary Fund, Dominique Strauss-Kahn, their misdemeanors can no longer easily escape the probing eye of the law or of the media.

It is in this new spirit that abuse of children by the clergy has lately been exposed in many parts of the world, particularly where people feel bold enough to challenge the authority of the 'pious' and of the custodians of the law and the public morality.

This shift in the moral assertions is not to everyone's liking, nor can it possibly be. Those who believed that the rule of law in fact meant the law of the rulers, who were thus above the law because of their political clout, social status, cultural trust, or wealth, can hardly feel very happy to face a jury of the common folk.

Does this mean that we are beginning to re-define our 'humanity'? Time and again we are discovering that each one of us, whatever our station in life, now has an inner world that is as deep as the ocean, and like the ocean, we are conscious only of what is on the surface. How some warm stream or cold current flows through the deeps and makes the surface of the ocean turbulent still remains a great mystery. When this turbulence brings disaster, we agonize about the nature of evil, and wonder about its source. 'What went wrong?' we ask. 'How could it be set right?' Yet what we observe, or even care to observe, is only on the surface. The complex phenomenon in the depths of consciousness remains hidden from the eyes of the world. Who can construct a profile of Thomas Jefferson with all the seeming contradictions and hypocrisies? How is one to know what monsters dwelt in Col. Russell Williams, and why?

Perhaps we are discovering for the first time that evil – destructive thoughts, gestures and acts against others – is not confined to any one particular group of people; it lurks in each one of us in a strange variety of ways. None of us can really claim to be a 'citizen above suspicion'. Yet the probing eye of the law and public opinion, even when it dares to see, sees but so little, and it sees without compassion and without forgiveness. And perhaps with little understanding.

Once we believed our own 'conscience', or the all-seeing and all-knowing eye of God, to be the real probing eye of our thoughts and actions. But for many that eye has shut itself and we wallow in our own secrets, and in our own guilt and uncertainties.

All we know is that all of us are all too 'human': somewhat heroic sometimes, but invariably with feet of clay.

❖

PART IV

A New Banquet Table

If we had a keen vision and feeling
of all ordinary human life,
it would be like hearing the grass grow
and the squirrel's heart beat,
and we should die of that roar which lies
on the other side of silence.

— George Eliot (1819-80), *Middlemarch*

22

A Squirrel's Heartbeat

Many years ago, after a day of torrential rains, as I was walking with a friend on a beautiful winding road in the mountains in Thailand, I saw a rather ugly looking worm crawl out of mud and soon disappear into a hole. This sharp contrast between a beautiful landscape and an ugly creature momentarily jarred my senses. How is it possible, I wondered, to find beauty and ugliness so intimately juxtaposed to each other? "Why would God create such an ugly creature?" I blurted out to my friend. "What could possibly be its purpose in life?"

These are harsh judgments on the universe; one should make them, if at all, with great peril to oneself. With disarming compassion, that is sometimes a woman's alone, my friend mused: "Perhaps that creature is asking the same question of us…. What is the purpose of this thing called man on this earth?"

People who have devoted a life time to studying these things – such as Professor Edward O. Wilson of Harvard – tell us that there are over one thousand trillion ants in the world – compared to 7 billion humans – and they weigh almost as much as the entire human population. They turn more soil than earthworms; catch more harmful insects than birds, and serve as the cemetery squads to remove more than ninety percent of the dead insects and other small animals. Without these ants, Wilson argues, the earth could not survive as a living planet, whereas it could very well without humans.

Over the centuries and millennia, humans have brought under their control, and have exploited, mutilated or manipulated almost all large animals on earth, in water and in air. Before the cunning, ingenuity and technology of man, even a whale, or an elephant, or a lion is desperately ill-equipped by nature to survive. For hundreds of millions of years, in the face of many predators, these and other creatures on earth have evolved to survive and flourish. But with the arrival of man on this planet, all forces of evolution have gone astray; many large and small creatures have fallen prey to man's exploits, and now are relentlessly beholden to his mercy. "What kind of creature is man?' they must ask and bemoan. "What might be his purpose on this earth?"

In the long and complex evolutionary history of life, there is nothing that has prepared vast wondrous creatures of this fragile planet, which some call Gaia – 'The Living Goddess' – to counter the bullets that fly from the helicopters overhead, or the formidable soup of industrial chemicals that flows in the rivers, lakes and oceans. Nothing, indeed nothing in their bones or in their genes, or in the synaptic connection of neurons in their brains, seems to have equipped them to camouflage themselves against man. They may weep – as elephants and others are known to do; they may cry and whimper, they may famish and go extinct, but they can survive, if at all, only at the pleasure and mercy of man, in his zealously guarded fiefdom.

One can stand in awe before the glories and wonders of man: "*What a piece of work is a man, how noble in reason, how infinite in faculties, in form and moving, how express and admirable in action, how like an angel in apprehension, how like a god!"* Why on earth, for what inscrutable purpose of God, has man been endowed with such disproportionate powers of cunning and brain over all others on the planet? What special path did this creature, called man, follow in its evolutionary history? Or, in what God's name did he chart a pathway so very different from all others on earth that he can bend so many, including his own kind, so wantonly to his will and to his whims?

I sometimes think that unless this overwhelming power of man is tempered by compassion and generosity of heart – not by any evolutionary twists and turns but of his own making – man's kingdom on earth must surely come to an end, as it has for numerous tyrants that have run rampant in human history itself. Man's end, however, will not mean the end of the earth, but only of his brutal, indulgent and unheeding kingdom on earth. Would ants, earthworms and beetles prevail then, so very little differentiated as they are in their cunning from the others? I wonder if these lowly creatures are not all lying low and are waiting just round the corner for some grand catastrophe to happen. Big boys of the animal kingdom may all be seized and controlled, but there is a great underground resistance building up, ready to rise in defiance.

They are a grand army of miniscule soldiers, and they know all the nooks and crannies, caves and gullies of this earth where no man has ever dared to go. The animal kingdom is far more complex and varied than what it appears on the surface, even to a highly trained eye: of some one and a half million species on earth, at least a million are of insects, most of them less than one cm in size, and all of them, forever on a relentless long march of their own. British geneticist J.B.S. Haldane (1892-1964), an ardent socialist and an atheist, when asked by theologians what he could infer about the mind of the Creator from observing the creation, responded: "An inordinate fondness for beetles."
Fondness indeed!

There are some 350,000 known species of beetles on the earth, far more than of any other animal. Marc Zabludoff, in his book *Beetles*, calls the earth 'A Beetle's Planet'; they are here, there, everywhere, on every continent, in every climate. "They may be the true masters of our planet," he says. They outnumber all the mammals, reptiles, the birds and the fish put together. They eat everything, there is nothing they don't use, reuse or recycle; for the ancient Egyptians, the actions of the scrab dung beetle were similar to those of sun god Ra; like the sun, it emerged regularly and unfailingly, and yet from nowhere in particular, and did wonders; for them, it became a symbol of transformation, renewal and resurrection.

On the other hand, there is the devastating virulence of the bark beetles in the conifer forests, as detailed by Andrew Nikiforuk in his book, *Empire of the Beetle;* for over a hundred million years, the family of the scolytids have fiercely attacked over-mature trees and have wiped them out.

Yet, unknown to humans, and much before their arrival on the planet, for tens of millions of years, the ants, earthworms and the beetles have been the real keepers of life on earth. Even deeper still, at the microbial level, first peered at by Dutch microbiologist, Antonie van Leeuwenhoek (1632-1723) through his diminutive early microscopes in the 17th century, bacteria, archaea and viruses make up most of the complex web of life on earth. Ours is, in fact, a microbial planet, incredibly rich and deep, and the overwhelming majority of it is utterly unconcerned with vertebrates, let alone humans.

Does anyone know what and who shall inherit the earth!

One of the most inconsequential and annoying creatures we encounter is the mosquito. I have heard some people refer to the poor and the marginalized as 'mosquitoes' or 'insects'; they have no worth; they are of little consequence; they are bothersome; their lives have no meaning. Thus dismissed, such people internalize this self-image – a life of no worth.

"My life is like of insects and mosquitoes," my mother would sometimes say in the later years of her life. "Why did God create me?" she would cry in utter desperation. We, her children, didn't hear her cries. We thought nothing of them. Yet at the same time, we young intellectuals often debated with utmost earnestness, about the theory of self, about Vedantism, and sometimes about Schopenhauer's ideas about will, or Jean-Paul Sartre's *Being and Nothingness*. Here was a world of ideas and philosophy that was so grand and eloquent that it left no room for the mundane and the inarticulate. It left no room for the insects and the worms of the earth.

Why is it that so often our engagement with thought or art makes us blind to the tumultuous grandeur of what is right before our eyes? We seek meaning in our lives, and bemoan the emptiness that haunts us. Yet we deny the meaning of others' lives, that others too may have some thwarted dreams under their drooping eyelids which may lie there like sleeping volcanoes. Mary Ann Evans, better known as George Eliot (1819-1880), a woman writer in Victorian England, and a contemporary of Karl Marx and Charles Dickens, observed with poignancy the lives of the ordinary women and their longings at a time when few rural women had a voice or a face of their own. "If we had a keen vision and feeling of all ordinary human life," Eliot wrote, "it would be like hearing the grass grow and the squirrel's heart beat, and we should die of that roar which lies on the other side of silence."

Squirrel's heartbeat! Not once did we feel that the heart of my mother beat with the same thump as that of others, as of my father, as of that galvanizing announcer on the All India Radio of my youth, or of singing and dancing girls on the silver screen. Never did we believe that our mother had any dreams beyond cleaning and cooking and producing children, and more cleaning and cooking, and more children. We didn't imagine that she was once young, or even perhaps beautiful. Could she have once, as part of some wanton youthful folly, caused a flutter or two in the heart of a passing young lad? I – we – couldn't imagine; we didn't *see* her, beyond a servile, illiterate, uncultured, uncivilized woman, who had not learned to accept her lowly status with gratitude.

Today, many years after her death, I speak of ants, earthworms and beetles, and how they nourish the earth, and how without these lowly, seemingly inconsequential creatures Mother Earth could not go on. But growing up, I could not tell my mother, nor did I know myself that like the ants and the earthworms, she turned the soil of our lives so we could flourish. That we could not go on without her; in fact, nothing could go on. Today I wish to shout from rooftops that *Being and Nothingness* is not about some abstract, and abstracted, philosopher in a Parisian café. If the great classic is to have any meaning at all, it has to be about my mother too. It has to be about her.

I could not tell this to her, for I didn't know it myself. I didn't know that if philosophy was to have any meaning at all it must speak to the very ground of life – with all its seeming ordinariness. If it shuts itself off in some hermitically-sealed coffin of arguments and counter-arguments, it becomes sterile and lifeless. It smells of arrogance. It becomes decadent and meaningless.

In Federico Fellini's remarkable film *La Strada* (1954), one sees this in the life of a half-wit woman, Gelsomina. In the poverty of post-war Italy, she has been sold off for a few pennies to 'an artiste', Zampano, who uses her to attract a crowd for his travelling show in which he breaks a thick iron chain with his puffed-up chest.

As a strong and rough man, the artiste Zampano surely knows how to break thick iron chains, but he has little idea how to break the chains that bind his heart and soul. He treats Gelsomina cruelly; he humiliates her and derides her. Despite her child-like beauty with which she sees and ennobles one and all around her, she is made to feel that her life has no meaning; she is a mere speck of dust on the virile vanity of the artiste. Until one day she encounters 'The Fool' in a travelling circus who tells her that even the lowliest pebble has a purpose in the universe. "And if it didn't, the whole universe will be meaningless."

What then is my purpose, the half-wit Gelsomina wonders? Your purpose, the Fool tells her, is to love this insensitive brute, this artiste, so that one day he could break his chains and walk through the gates of redemption.

How I wish I could console my mother with some such words. "My life is like that of worms and insects," she would cry. "It has no meaning." With all my erudite learning, I had no words to console her, for, in truth, I didn't hear her cries.

Over the years and decades, I have known such cries all over the world; sometimes they are audible, sometimes they are no more than a whimper. And I have heard myself cry too when I have felt that I was nobody. Or when my heart had turned cold, and I could not hear the heartbeat of any squirrel anywhere.

Many years ago, on a rain-drenched afternoon, on a remote hill, in a country not my own, I had made a casual remark about the ugliness of an unknown worm that had slithered into a hole in the earth. It was not a place where I or my kind could go. But it was a place – a home – where abundant life must live abundantly, with its own kind of loving, and its own kind of meaning. Any judgement on this life, as I had so brashly made, was in fact a judgement on the universe. Man has made such harsh and ignorant judgements time and again, even against his own kind, even against one's own mother, as I did, and has thus destroyed many a home and many lives. That slithering ugly worm taught me, as much as any great teacher can teach, that life has innumerable expressions; sometimes their purpose eludes us humans, but in God's name, or in the name of the mystery of the universe, man can only be a silent witness to them, and not their judge.

"I never saw an ugly thing in my life," eighteenth-century English artist John Constable once declared. "Let the form of an object be what it may – light, shade, and perspective will always make it beautiful."

Is it possible that with a different perspective, in a different light, in the eyes of a true witness, all things become beautiful?

Beautiful and full of grace?

Is it possible?

❖

23

At the Banquet Table of History

George Orwell once observed: "Who controls the past controls the future. Who controls the present controls the past." History, he meant, is told by the victors, in which the vanquished, the marginalized and the downtrodden have little, if any, place. And then what is told as history – in schools, churches, books, and songs – assumes a life of its own; it rings unquestionably true to the succeeding generations.

History has indeed been a grand banquet table on which only the ruling classes, the elites, the Brahmins, and only men sat and gorged themselves. That is how so much history of the world has been Eurocentric, in which Europe has been the great centre of civilization, and its great heroes have been almost always men. History is 'his story', many feminists have railed, and not 'her story'.

One of the hallmarks of life in the past half century has been an attempt – many attempts in fact – to re-tell history from different perspectives. Increasingly there are many more narratives of the human story, all vying for our attention. Today everyone – Asians and Africans, labourers and artisans, women and *dalits*, the handicapped and the weak – is clamouring for a place at the banquet table of history. "I too am here," they shout. Like Willy Loman in Arthur Miller's much celebrated play, *Death of a Salesman*, desperate voices are arguing: "I don't say he's a great man.... But he's a human being, and a terrible thing is happening to him. So attention must be paid. He's not to be allowed to fall in his grave like an old dog. Attention, attention must finally be paid to such a person."

Indeed, attention must be paid; the value of each life – of every child, of every woman and man – must be recognized. This is the great desperate cry of our times. For far too long this cry has been snubbed and muzzled. Now its time has come to be heard, and heard succinctly, and with all the moral principles of social justice.

More than two hundred years ago, in 1798, Rev. Thomas Malthus was so troubled by the rising population of Britain that he wrote a revolutionary book, *An Essay on the Principle of Population.* He cautioned against the exponential increase in the human population in contrast to the limited increase in food supply. And he postulated that "crime, disease, war, and vice" are necessary checks on population. At that time, he could not envision – in fact, no one could – that a few decades later there would be grand technological advances to control birth, and increase agricultural produce dramatically. Or that there would be a large societal change in the desire and need for large families. And so, all an imperial power could imagine in the nineteenth century was to plunder others' resources, occupy their lands, and engage in wars of conquest and empire, and feel morally justified by the 'laws of nature' for doing all that.

Between the first census in 1801 and another a hundred years later, in 1901, the population of Britain had increased almost 400 percent, even as millions had migrated to other lands, including Canada. Yet, at the height of its imperialist expansion, Britain was in a sorry state at home. No one has portrayed that state, particularly of the children, with greater poignancy than the novelist Charles Dickens. In his *Great Expectations*, he writes: "In this little world in which children have their existence there is nothing so finely perceived and finely felt as injustice." It was during this period that over a hundred thousand poor and orphaned 'Home Children' were shipped off from Britain to Canada and Australia to work under the most horrific conditions.

The idea of justice goes back into human history for millennia, but we still differ greatly about what we understand justice to be. Injustice, on the other hand, seems more obvious. Many argue that to reduce injustice in the world is itself a noble and humane goal. All religious traditions have encouraged their followers to practice charity and philanthropy. "But charity is not social justice," many argue with great passion; its moral principles must be, they insist, beyond and above the goodwill of the individuals.

The human story has been so riven with violence and wars, and with exploitation, abuse and humiliation of so many – women, children, the poor, the weak, the handicapped – and over such a long period, that instances of compassion, kindness or even tolerance of others as human, are noticeably few. Yet we humans have survived, and have in fact multiplied in such abundance, that this 'achievement' borders on the miraculous. Why and how has it been so? Why have the homo-sapiens – the 'wise ones' – been so addicted to violence, domination of one another, and the abuse of the weak and the meek? At last we are beginning to realise, however slowly, that slavery is abominable, that freedom of thought and life can be cherished by all, that others too have a touch of humanity in them, that we live in an inter-dependent world and we all share one ecologically integrated biosphere, and that beyond the colours and creeds and castes and races there is the same rhythm to the human heart.

Though radical and revolutionary, this ethic did not emerge from some moral void. It emerged, in the age of science, from the ashes in concentration camps in the Holocaust that saw murder of six million men, women and children because they were Jews or Gypsies, or Jehovah's Witnesses, or homosexuals.

This new ethic is grudgingly emerging from hundreds of years of petty national and religious wars on many continents; it emerged from the genocides by imperialist European powers of native indigenous people in the Americas; it emerged from the brutality and inhumanity of slavery and slave trades over centuries; from the oppression and suffering of women, not only during wars, but in everyday domestic and public life as well; from the pain and scars of serfs, indentured and bonded labour, of landless peasantry and of industrial workers; from the neglect and exploitation of children, and from a thousand other brutalities that have been heaped on the marginalized, the poor and the defenceless for millennia.

Above all, it is emerging from unexpected turns in historical events, and from the depths of human conscience, from the sympathy and concern of men and women of imagination and goodwill, and from our inherent desire, however faint or ignored, to live by some 'moral instinct'.

Such an ethic was long time in the coming, and it has far from arrived, but it promises to empower men and women more than ever before in human history. And it is not an empty promise, like that of heaven or hell, or of after-life, so vociferously made by many religions for centuries. It is a promise of the here and now. "It is *my* body, and *my* mind!" this new ethic claims. "I must have complete sovereignty over them, without fear and without interference by anyone." "What makes one race superior to another?" this new ethic questions. "Who decides homosexuality is unnatural and ungodly?" "How is it determined that women are not *persons*, and are incapable of this or that?" "Are children the property of their parents, or are the parents merely their guardians?" "Why not treat the handicapped with dignity and respect?" "Who defines what is it to be human?"

Perhaps it could be said that all those who had been denied a place on the banquet table of history for millennia are now at least clamouring for a little place, and fewer and fewer feel that they have the moral superiority to keep them away from it forever.

Could Rev. Malthus see any of this two hundred years ago? Could anyone see this even after the end of WWII? The fact is, even the most astute political observers could not see the end of the Soviet Union in 1990, or the dismantling of the Berlin Wall in 1989, even one year before they happened.

Could anyone see the Quiet Revolution coming in Catholic Quebec, where the family size of 8-10 children was the norm until 1950s, ushering in the lowest birthrate in the world in just a few short decades? Now in many countries, with little or no immigration, there is a danger of shrinking populations and increasing longevity. Such is the case in Japan, Italy, and Spain. In other countries with large immigration, such as France, Britain, Holland, Canada and the USA, the older generations are often apprehensive about the new face and colour of their country and society; they talk of 'Islamization of Europe' or of 'Eurostan'; they talk of the end of the world, certain that the new immigrants – particularly Muslims – will continue to breed like pigs.

For many, it is possible to see the changes coming in their own society, and yet they view other societies as fossilized, incapable of any change. After all, what other country in human history, except China, has introduced, and largely accepted, the one-child family policy, with its far-reaching consequences?

When young men and women are not seen as mere fodder for the war machine, to be sacrificed and killed in arrogant and reckless wars, each person's life assumes a certain sanctity. Does not every life, in fact, have an undeniable sanctity for those who are touched by it, as a father, or a mother, or a brother, or a friend? Or by human empathy?

There are now over 7000 million people in the world; each one of us is charting our own journey, with our own constraints and our own freedoms.

We need to walk gingerly on the planet with what we do know, but we can lay the map out only so far to what might possibly unfold, before it gets rendered obsolete by the opening of new roads and new paths.

I believe that when it comes to the human story, no one, however insightful, and however equipped with vast data and forecasting power of computers, can see very far. With 7,000 million people, and with at least 7,000 million degrees of freedom, there is only one certainty about this story: the paths we chart, the walls we break, the holes we peek through, the roads we walk on, and the dreams we dream, are far from certain. With such a vast number of degrees of freedom, one can only say that we live in a most uncertain and most creative universe.

❖

24

The Census and The Conscience

In his office at the Institute of Advanced Studies in Princeton, for many years Albert Einstein had a sign hanging that read: "Not everything that *counts* can be counted, and not everything that can be counted *counts.*"

I believe Einstein, in his own inimitable way, was trying to caution that not everything that can be counted and measured is of real worth; as well, not all things of real worth can be measured. Over the years, I must confess, I have developed a certain disdain for those, who in the name of science, collect the most trivial data, often at great public cost, that contributes little to developing any picture of the underlying phenomenon. Data, for them, becomes an irrefutable substitute for perception; often they are blinded by it. "It is the spirit of the age," author Gore Vidal once lamented, "that any fact, no matter how suspect, is superior to any imaginative exercise, no matter how true."

Yet, is every perception that is not in accordance with facts, or is contrary to them, mere fiction? Is it to be dismissed as misleading? In February 2011, India concluded the largest data-collecting exercise in human history. What it set out to gather, and what was revealed by it was quite mind-boggling: the 2011 Census of India – the 15th such census since first conducted by the British under Queen Victoria in 1872 – was the largest of its kind anywhere, covering some 240 million households across the length and breadth of this vast and heavily populated land.

India makes no sense without her massive numbers; everything here is in millions, and hundreds of millions. Almost every state of India is bigger than most countries in the world; some are even three of four times larger. Number of malnourished adolescents is in hundreds of millions. The number of people in middle class is in hundreds of millions. The number of rich and super rich is in tens of millions. The number of people riding trains – many hanging on the outside – is in millions. The number of mobile phones is in hundreds of millions, as is the number of cars.

The same is true of money in India. Money that is stolen by politicians, bureaucrats, judges and contractors is in hundreds of millions. Food that is wasted in storage houses, or in transportation, or at gala weddings, is in hundreds of millions of tons.

This teeming, encroaching, indifferent mass of humanity can numb one's senses and sensibilities; it can even accentuate one's self-centeredness. It hardens your defences; it insulates you from those who are not part of your tribe. It forces you to close your eyes. To find some private space of your own, you may be drawn into meditation. Many are drawn into some religious mumbo jumbo, for some solace, for some meaning, for peace. Temples, gurudwaras and mosques here are filled with devotees; their coffers are filled with hundreds of millions of rupees and tons of gold. Hospitals are famished for medicines and equipment. Thousands of rural schools are devoid of blackboards and teachers. Libraries are non-existent. But god-men live in luxury.

There is a great deal that the new census set out to reveal: hundreds of millions who were disabled were counted for the first time; tens of millions who are 'transgender' were counted for the first time. The number of mobile phones and internet connections; the number of divorcees and 'live-ins'; the number of the aged and the infirm; the number of languages spoken and cast groups; religious affiliations and atheists and the unaffiliated, were all meticulously recorded.

There was a great deal that was revealed about 1200 million people; the information will be necessary, even crucial, for India's transformation.

For over four thousand years, countries everywhere have been collecting information and data about their people, for recruitment into armies, for taxation, even for care and development sometimes. In the early Pharaonic period, between 3340 and 3050 BC, Egypt collected data about its people. Two thousand years ago, in 2 AD, China recorded 57.67 million people in 12 million households in the Han Dynasty.

But it was the revolutionary book, *An Essay on the Principle of Population* by Thomas Malthus that started a more modern concern, in 1801, about the size and distribution of population, as outlined in numerous debates in the British Parliament:

- the intimate knowledge of any country must form the rational basis of legislation and diplomacy;
- an industrious population is the basic power and resource of any nation, and therefore its size needs to be known;
- the number of men who were required for conscription to the militia in different areas should reflect the area's population;
- there were defence reasons for wanting to know the number of seamen;
- the need to plan the production of corn and thus to know the number of people who had to be fed;
- a census would indicate the Government's intention to promote the public good and
- the life insurance industry would be stimulated by the results.

The immense amounts of data collected in the census, and the complex picture of India that emerged from it, was no doubt infinitely valuable to all who are engaged in the future of India as an economic, cultural and political entity. But the vision of that entity as a just and equitable society, caring for her famished children and of her teeming millions, or as a society riddled with petty and exploitative identities and interests, will surely come not only from a census but from a social conscience.

Such a social conscience seems to be languishing in India more today than perhaps ever before; certainly it is more lacking than I have known for decades. Many more millions – the young no less than the old – seem to be more self-centred, more exploitative of their fellow citizens and of the environment, more reckless in their ambitions, more ruthless in their gluttony, more convoluted in their ethics, and far, far more clever in their moral escapades. Old learned men and women now hark back to *dharma* and *karma* far more than they ever did before, quite certain that all solutions are to be found in the *Mahabharata*. However, their solemn words sound rather empty and shallow, and quite self-serving. Behind their punditry they wear a mask that one can barely pierce through. I too then wish to close my eyes and meditate on some slumbering God. But alas, it hardly makes me feel very peaceful.

Collecting data and gathering information are good things. But as the great scientist kept reminding everyone to his dying day, "not everything that counts can be counted." For that one must look for a different light, in a different sky.

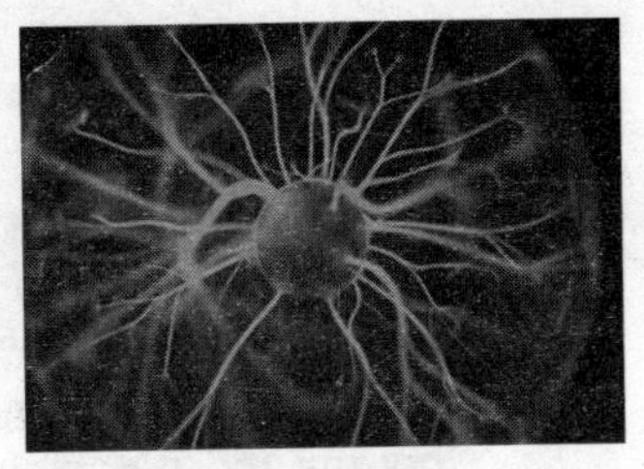

25

Celebration or Chaos?

The 2011 census in India, the largest in the human history, revealed a population of more than 1,200 million. In 1947, at the time of independence, and after her partition, India's population was some 350 million. That is an increase of more than 300 percent in 64 years. Overall, the numbers have been increasing in the whole world equally dramatically: from 3,000 million men, women and children in July 1959, to 4,000 million in April 1974, to 5,000 million on July 11, 1987, to 6,000 million on October 1999; more than a thousand million new beings added every 12-15 years. On August 26, 2011, it was declared that were 7,000 million of us on the earth. By mid-century, in 2050, we are likely to grow to 9,000 million.

Over two thousand years ago, in China, according to the census collected during the Han Dynasty in the Fall of 2 AD, there were 57.67 million people. Today that number is past 1300 million.

Despite all the wars and famines, plagues and devastations, human growth has been most prolific. It has been termed a 'Population Bomb'; many bemoan the vast numbers as disastrous for the ecological well-being of the planet. Yet, if one were to ask anyone – however impoverished or altruistic – if, for the sake of the earth, they – or their friends or families – should not have been born and had chosen not to burden the earth any more, the answer would be a resounding NO. The problem is surely with the others, never with oneself.

Is there ever anyone on the earth who believes that his own life has been a great burden on this fragile planet?

The vast numbers obviously mean more mouths to feed, more schools and more shopping plazas, more jobs and more homes, more wastes and more cars. More water. More energy. More needs. More expectations. More competition. More struggles. More and more and more of everything.

At the rate we and our needs are growing we would need more than one earth. Perhaps 4 or even 5. Yet who is ready to jump off the planet and declare, "That's enough! It has all been a big mistake"?

In the frantic metropolis of Delhi, in the midst of thousands of cars, three-wheelers, buses and trucks that drive every hour of the day and night, amongst a plethora of billboards, next to one that reads, "For That Fifth Avenue Look", there is one with Mahatma Gandhi's picture and his somewhat prophetic words spoken more than 80 years ago: "Earth provides enough to satisfy every man's *need*, but not every man's *greed.*"

I wonder how, and when, something is no longer a matter of greed; it simply becomes a need. When did a car become a need? One car? Two cars? Three cars? Since 1947, India's car population has probably increased by 2000 percent, possibly more. On the university campuses in India now, amongst its burgeoning educated elites, there are few bicycles to be seen. Only cars and motorbikes. Surely, the human burden on the earth cannot be calculated only by the number of people on it; it must also include what each person consumes, wastes, and destroys – 'the ecological footprints' each of us leaves on the earth.

Yet everywhere, the number of people in any nation is considered as a sure sign of underdevelopment, and the number of cars as a sign of development, and a proud symbol of an unrelenting juggernaut of progress. The former would inevitably, many argue, create a nightmare for the earth, and the latter is a dream of progress come true.

What a strange fallacy!

Despite all the devastations, why do human beings keep growing? "Humans are a cancer on earth," some have argued in desperation about the impact of human presence on the planet. For many ecologically-minded people, human presence, in such large numbers and with such foreboding technological prowess, is a threat to the survival of the entire biosphere. They bemoan that humans are an earth-devouring species.

I feel obliged to speak of humans in less belligerent terms. Human beings don't only have mouths to fill, they also have brains, probably the most complex and evolved expression of matter in the universe. Could this growing number of human beings also mean more intelligence, more freedom of thought, more consciousness, more creativity, more insights? More discoveries? More co-operation? Fewer wars?

Could they also mean more hope, and a different dream for the planet? A different world? With 7,000 million degrees of freedom, can a new world come into being?

A safe and reliable way of birth control, a pill, was discovered only sixty years ago, but that single invention brought about, for hundreds of millions of people, a new freedom from having large families; it ushered in a new definition of the role of women; it began to redefine our sense of morality. For many, it even meant questioning the role of God or the Church in their lives. What seemed inevitable and even part of the 'natural law' to such thinkers as Thomas Malthus in his revolutionary book, 'An Essay on the Principle of Population', in 1798, suddenly seemed far from it. Malthus wrote:

> Must it not then be acknowledged by an attentive examiner of the histories of mankind, that in every age and in every State in which man has existed, or does now exist
>
> That the increase of population is necessarily limited by the means of subsistence,
>
> That population does invariably increase when the means of subsistence increase, and,
>
> That the superior power of population is repressed, and the actual population kept equal to the means of subsistence, by misery and vice.

For Malthus, misery, vice and wars were nature's way of keeping a check on the rising human population. In 1798, could Malthus dream that a birth control pill, rather than wars and diseases, would change the human story so drastically? Indeed in many countries in the world – Japan is the foremost example of this – human population is decreasing so dramatically that it is a matter of concern for their survival as a viable nation.

Similarly, could we imagine a world in which physical travelling – requiring cars, aeroplanes, buses – would be a thing of the past, and we would become, instead, virtual travellers? Information, ideas, images, words, money, books, music, films and so much more now travel at the speed of light, virtually – all of which was unimaginable even fifty years ago. What would it be like if this could happen to 'things', to matter in general?

As with the pill, or mobile telephones, or the internet, one small turn in the journey, whether wrought by an invention or a social idea, and a whole new primordial forest of possibilities opens itself to one's view. Or equally, an unfathomable abyss.

This, I believe, is the grand phenomenon of man; at its heart sleeps the coiled *naag* of freedom. Freedom to dream a new world is our unique and great gift. With hundreds of millions of men, women and children, our core of consciousness has also multiplied a hundred million fold. So many degrees of freedom – hundreds and thousands of millions of them!

This multi-dimensional universe of human imagination would defy any Einstein. It can be modelled by no statistician or a computer. Its possibilities are infinite. "Now my own suspicion is," observed distinguished geneticist and an ardent atheist, J.B.S. Haldane (1892-1964), "that the universe is not only queerer than we suppose, but queerer than we *can* suppose."

And Haldane, I don't think, even postulated the role of consciousness, or even of human consciousness, in this incredible universe. As a geneticist, he could envision how an unexpected and 'queer' Book of Nature was being written even with a limited alphabet. How much grander would the epic of the universe be if one could see how it is imbued with consciousness, and with new and ennobling human consciousness, at its centre of gravity, at the core of its being.

In Q2C – 'Quantum to Cosmos' – Festival in December 2009 at the Perimeter Institute of Theoretical Physics, the venerable physicist Stephen Hawking declared to great applause: "My goal is simple. It is complete understanding of the universe, why it is as it is and why it exists at all."

The insights and discoveries of physicists in the past several decades have been revolutionary. Yet this goal, expressed without a touch of irony, seems to me rather unimaginative, and quite utterly and hopelessly unattainable, unless, of course, the universe he wants to understand completely is dead, unchanging, uncreative, and without consciousness. The wonder of the universe is that it is forever making itself anew, if not among the stars and the galaxies, but certainly here on earth where we experience the universe with our every breath. What we discover is that, despite its numerous laws and cycles, it is an unexpected and creative universe, full of surprises and full of new turns of events and thoughts.

I believe that so long as there is life in this universe, so long as a leaf moves in the wind, so long as water flows in a stream, so long as a thought stirs through a brain, so long as a man wonders and enquires, the universe will always be in a state of flux, always at the edge of the unknowable, never fully calculable. Ten thousand years, or a million years into the future, science may predict when a solar eclipse would occur, or a star may go cold. But no science can predict at what precise moment a child would utter what word, or what thoughts would fly through her mind. We live in a universe of infinite degrees of freedom. Even when it is often squashed, or is not exercised, freedom is what defines life forever. Freedom is the warp and weft of a living, creative universe that sometimes makes some sense but often remains utterly incomprehensible.

With 7000 million degrees of freedom, will there be a cosmic celebration or a cataclysmic chaos?

❖

26

Sharing a Saddle on the Back of a Camel

In 1492, when Christopher Columbus sailed from Spain to discover a westward sea route to India, his goal was to find converts for the Catholic Church, and slaves and gold for the Spanish monarchy. On his two subsequent voyages, he reported, to his surprise and, perhaps, slight disappointment, that "in these islands I have so far found no human monstrosities, as many expected, on the contrary, among all these peoples good looks are esteemed." These Indians, he assured the Spanish sovereigns, were "very well built, of very handsome bodies and very fine faces."

As well, he reported, "They are so ingenuous and free with all they have, that no one would believe it who has not seen it; of anything that they possess, if asked of them, they never say no; on the contrary, they invite you to share it and show as much love as if their hearts went with it."

Yet, over the centuries, the discovery of the 'New World' is a gruesome story of desecration and destruction of the indigenous people and their cultures by the Europeans. From the very beginning, as the Indians were subjected to 'cruel and horrible servitude,' and enslavement, the humanity of the Indians, and their potential equality in the sight of God, were disputed by the Spanish settlers. In some circles, it was being argued that if Indians were not fully human, how can they be converted into Christians. And if they were indeed human, how can they be treated so inhumanely. This debate raged on for more than 50 years after the discovery of the New World, and was finally settled in 1556 through a papal bull, though the cruel and horrible servitude of the Indians never subsided.

Unlike the Hindu scriptures, for instance, Bible seemed clear on the single origin and homogeneous descent of the whole human race. Since all men were descended from Adam and Eve, there was no place for the inferiority of genetic endowment. The differences were those of language and religion alone, and not in 'humanity' itself. In Europe, as the Church of Rome was losing millions of souls to the Protestant heresies, the New World suddenly offered, as though by divine providence, its countless pagans for a vast harvest of new believers. By 1540, despite all the debates in the Vatican, some six million American Indians had been baptised.

What abominable obscenities have been committed in the name of 'purity' – purity of blood, purity of races, purity of women, purity of thought! Over two thousand years ago, Plato presented a 'magnificent myth', the Myth of Metals, according to which 'Mother Earth' is fashioning people differently, some with gold, and others with silver or bronze. In *The Republic* he argued that human reproduction should be monitored; mates would be chosen by a 'marriage number' in which the qualities of an individual would be quantified and analyzed, and persons with high numbers will be allowed to procreate only with similarly high number. Uncertain about the process of inheritance, Plato did, however, acknowledge that 'gold soul' individuals could still produce 'bronze soul' children. If the myth were believed by the people, he surmised, it would prove to be self-fulfilling and would turn out to be true: alleged differences 'in the nature of things' would be confirmed by social practices and fiction would become fact. And then finally, the Guardians of the Republic themselves would be convinced that the 'magnificent myth' is true.

Similar myths have been perpetuated in the Manusmriti and the Vedas about the *sudras*, the outcastes and the 'untouchables' in the Hindu society: "If the sudra listens to the

Veda with the intention of committing it to memory, his ears should be filled with molten lead. If he utters the Veda, then his tongue should be cut off; if he has mastered the Veda his body should be cut to pieces," says the Manusmriti. The centuries-old divisions of Hindu society into castes, and the condemnation of hundreds of millions into untouchability and apartheid, are as shameful features of a culture as racial segregations and brutalities have been in America. That they linger on in India, even in the 21st century – and encompass not only Hindu social structures but also those of Sikhs, Christians, Muslims and Parsis – speak of their hydra-like power with multiple heads.

In 1981, on a study trip to Calcutta with a group of my students from Canada, on a visit to St. Paul's Cathedral on Easter, we discovered a whispering campaign in the congregation against the newly-appointed Bishop from the 'untouchable' caste; to the *bhadralok*, the gentry, of the Church, the Harijan Bishop from the South was clearly unacceptable.

Of course, the Nazis, under Hitler, made their ideology of the superiority of Aryan race into a new abomination. Insidiously supported by many in occupied countries such as France, Norway, Hungry and others, to propagate their ideology of pure race of Aryans, they wiped out millions of Jews and others they deemed undesirable; they also created Lebensborn homes in many countries, where local women could produce pure Aryan children with German soldiers and thus propagate the ideals of the Third Reich.

Purity - of blood, race, culture, ancestry, thought or even of mathematics and science – is a delusion. It is the presence of the other, the impurities, which render resilience and tenacity to everything. Pure gold – 24-carat gold – is too soft; when it is mixed with something else – silver, nickel or palladium – it gains resilience and colour. When people of different ethnicities, cultures, religions and histories mix, the great tree of life grows and it branches out in unexpected new directions and rises to new heights. This is the brave new world of which my mother received only a disconcerting inkling in her dying days. But for the new youth, this new world seems as natural as the divided world was for us and our elders.

When a people speak of 'khalsa', the 'pure ones', or of Pakistan, the 'land of the pure', they only propagate delusional chauvinism. Sadly, such assertions are to be heard everywhere: the Québécois term 'pure laine', literally meaning *pure wool* (and often interpreted as *true blue* or *dyed-in-the-wool*), is a politically and culturally-charged phrase referring to the people having original and pure ancestry of the French-Canadians. With Quebec now as a multicultural society, one needs to ask afresh: Who is a true Québécois? When will the idea of 'pure laine' become obsolete?

For millennia we have been obsessed with stultifying notions of purity; how liberating it is to discover that it is at the interfaces of many seemingly irreconcilable forces, such as the *aradhnarishwara,* half-man and half-woman, that real creation takes place, and life blossoms in its most sumptuous fecundity. Not arrogant notions of purity but generosity of heart will have to be our new watchword: generosity, acceptance, diversity, celebration. The true expression of love, it needs to be said, is to share a saddle, far more than to share a bed.

Sharing a saddle on the back of a camel, as it trudges along on a long journey in the deserts of life, is the only true and pure act of love.

❖

27

In the Skin of Another

In 1959, a white man in USA, John Howard Griffin, wanted to know what it feels like being a black man in racially-charged southern states of America. With the help of a dermatologist, he changed his outer self, and assumed the persona of a black man, and then travelled in many cities in the south to experience the fate of blacks first hand. On the basis of these experiences, in 1961, Griffin wrote his book, *Black Like Me*.

Griffin's remarkable transformation was baffling. In his own words:

> *The completeness of this transformation appalled me. It was unlike anything I had imagined. I became two men, the observing one and the one who panicked, who felt Negroid even to the depths of my entrails. I felt the beginnings of great loneliness, not because I was a Negro but because the man I had been, the self I knew, was hidden in the flesh of another.*
>
> ...
>
> *The transformation was total and shocking. I had expected to see myself disguised, but this was something else. I was imprisoned in the flesh of an utter stranger, an unsympathetic one with whom I felt no kinship. All traces of the John Griffin I had were wiped from existence. Even the senses underwent a change so profound it filled me with distress. I looked into the mirror and saw nothing of the white John Griffin's past.*
>
> ....

I felt strangely sad to leave the world of the Negro after having shared it so long – almost as though I were fleeing my share of his pain and heartache.... I felt their arms around my neck, their hugs and the marvellous jubilation of reunion.

Griffin's vivid exposé of the hatred against the blacks in the USA, "of being treated like a tenth-class citizen," drew much admiration from many, but for people of his town and state, he was an impostor; he was vilified and threatened with death. He had betrayed his own kind; he was a traitor.

In 1976 Richard Dawkins wrote a book that has since become a landmark for what it proposed and also for its title, *The Selfish Gene*. When I first read the book, I found his ideas about the 'self' baffling; he seemed to treat a gene, like much else, as an object within its own skin, with no communication with others. Like a rubber ball. A gene is hardly like that, I demurred; how can it be 'selfish'? For me then, and ever since, anything living is always in communication with others, however few, and however meagrely. And this process keeps making the living ever new. In other words, a self is fluid, ever evolving, always in the making.

Our suffering – human suffering, not the suffering of the gene, about which I can't speculate – begins when the self that defines us, psychologically and spiritually, closes on itself, stopping all flow. That is when our words and acts become only 'self-serving'; we become self-centred. No one else matters to us. Feelings of love and concern do not flow through us, and out of us. All of us occasionally feel such moments, but is it our natural state of being, or a state we aspire to as evolved beings? When our self is expanding, and it is breathing out to others – to our children, to our brothers and sisters, to our ancestors, to our fellow-travellers and to strangers, to life forms that crawl and fly or slither into holes and crevices, to ideas and passions that move and touch us, to music that stirs us – what we call "I" takes flight; through empathy, we assume the skin of another. We walk in the shoes of another. We transform our self.

It is thus that the 'self' we sometimes so carefully guard

from the onslaught of others, or yearn to be noticed and known for what it is, is always far more than what it seems on the face of it. Above all else, it finds fulfillment in reflecting itself in others, in bonding with others, in reaching out to others. In touching others.

In his book, *Souls on Fire: Portraits and Legends of Hassidic Masters*, Elie Wiesel narrates a story:

> A disciple tells the Kotzker his woes: "I came from Rizhin. There everything is simple, everything is clear. I prayed, and I knew I was praying; I studied and I knew I was studying. Here in Kotzk everything is mixed up; confused; I suffer from it, Rabbe. Terribly. I am lost. Please help me so I can pray and study as before. Please help me to stop suffering."
>
> The Rabbe peers at his tearful disciple and asks: "And who ever told you that he is interested in your studies and your prayers? And what if he preferred your tears and your suffering?"

I have often pondered on this story and have wondered what it could mean. I know we are all obsessed with our self; that is absolutely necessary, and natural. "If I am not for myself, who shall be for me?" ask the Jewish Talmudic scriptures. But the 'self' that defines our pain, our pleasure, our obsessions, and our sense of worth, or the purpose of our life or of our being, is so multi-faceted that it sometimes surprises and baffles us. What this 'self' is, and where its boundaries are – where does it start and where does it end and become 'non-self' – is, in numerous ways, the most fundamental biological, philosophical and spiritual question.

Genes may be selfish, though I doubt that very much. But if human beings were selfish, we would have stumbled in the first few steps of our long and arduous journey across millions of years. It is through cooperation, through care of our children way past their need for care, through life-long affections and bonds, through passing on the light of our learning and skills and knowledge to the generations to come, through sharing and teaching, and through our dreams and our memories that we weave a universe that stretches across oceans and mountains, across eons of desecrations, through ice ages and wind-swept barrenness. The real lesson that all of us must learn is that we are more than our self; or that our self is more than itself.

Returning to the words from the Talmudic scriptures: “If I am not for myself, who shall be for me?” we come to the second line: “If I am only for myself, what am I?”

It is in response to this second question: “What am I?” that we are called upon to become – each one of us – the practitioners of the greatest art there is: life itself. Life as Art. Before this art, all else is a mere shadow, a phantom attached to a phantom body. “An artist is not a special kind of man,” comments art historian Ananda Coomarswamy. “Every man is a special kind of artist.” In this act of creation, each man becomes part of the Creator – a creator that is compassionate and caring. And not selfish at all!

How does this fit into a crude understanding of the idea of ‘survival of the fittest’ that social Darwinists so flaunted for decades, and still do? How often the weak and the marginalized are dismissed as being ‘naturally’ inferior, incapable of learning, or worse, as brutes and savages. Yet the same marginalized people, given a chance to break through that iron wall of prejudice and subjugation, have created wonders.

We pride ourselves as rational beings, and imagine that the ‘Age of Enlightenment’ in Europe had ushered in reason and pushed out prejudice and obscurantism. Yet our history proves again and again that we humans have a great instinct for rationalizing; we justify our biases and prejudices with clever reasoning. We are not only ‘rational beings’, we are also ‘rationalizing beings’.

There is a dire need for self preservation in nature but there are also magnificent expressions of cooperation, mutual help, even of altruism, where one organism makes sacrifice of its personal interest for the sake of others. Mammals can act with great altruism on behalf of their offspring, as penguins do. Think of the reciprocal benefits that flowers and bees bring to each other through the process of pollination. This cooperation – or mutualism or symbiosis – increases the chances of the genes of each of them surviving and flourishing.

In his famous essay on *Evolution and Ethics*, in 1893, Thomas Huxley explores the moral messages that nature might encode for humans:

> The practice of that which is ethically best – what we call goodness or virtue – involves a course of conduct which, in all respects, is opposed to that which leads to success in the cosmic struggle for existence. In place of ruthless self-assertion it demands self-restraint; in place of thrusting aside, or treading down, all competitors, it requires that the individual shall not merely respect, but shall help his fellows...It repudiates the gladiatorial theory of existence...Laws and moral precepts are directed to the end of curbing the cosmic process.

The human story too, with all its cruel battles and skirmishes, is marked by extraordinary family cooperation, tribal affiliation, personal sacrifices and altruism. Indeed, we would have perished long ago in every ice age and in every drought, or been lost in wilderness and snow storms, were it not for a stretched-out hand, a voice of consolation, or the warmth of an embrace. We have persevered as humans despite a million frailties of the body and mind. It is because somewhere in the marrow of our bones, a message has travelled through vast passages of time that we cannot take this arduous and unknown journey across the continents, over the oceans, through the unlit sky, or even through the vast stretches of imagination that illuminate our being, without holding each others' hands, and without some guiding light that comes from within each one of us, and also shines amongst us.

An urge for justice, for lawfulness, for life to prosper rather to be diminished, and for birth and rebirth are all part of our biological, cultural and spiritual heritage. 'Are we wired for ethics?' some people wonder, as they search for an imprint in the DNA molecule for an evolutionary advantage for being ethical. 'Are we born sinful? Or, 'in the image of God?' some others wonder, and mourn all the pitfalls we have encountered and have fallen into. Others invoke *karuna*, compassion and forgiveness as our guiding light in the great storm.

There is no single syllable, or *mantra*, or theory that captures the origin or need for ethics in the human story. Nor can we know, once for all, what it means to be ethical under all circumstances. Our conscience may tell us a little; the words that have travelled over the centuries from the mouths of sages or from the books of wisdom may guide us on. The laws and practices that we have devised may serve as our boundaries. Yet, there will be moments – there will always be many moments in our lives – when we will agonize and wonder, 'What is the right thing to do?' All one can hope, invoking the illuminating words of poet W.H. Auden, is to 'stand at the window/ As the tears scald and start ... [to] love your crooked neighbour with your crooked heart."

28

The Hands of Barbarians

In his iconic film, *Modern Times*, made during the Great Depression, Charlie Chaplin shows, in his own inimitable way, how the pace and the demands of the machine overwhelm the humans, who become, inevitably, 'cogs in the machine'.

Time and again we experience the power of machines on our lives: of cars and aeroplanes, of weapons of destruction and instruments of torture, of internet and mobile phones. It is as though these machines have assumed a life of their own, and have taken over our lives in a million different and inscrutable ways.

In 1912, just two years before the outbreak of the First World War, and in the early years of the invention of aeroplanes, the New York Times wrote in an editorial that no civilized nation could imagine using such machines for attacking the others from the air. Less than a year later, it had become apparent to all that no 'civilized' nation could afford *not to use* these machines to bombard the others.

Technology has its own imperatives. It is thus that the triumph of science in the 20th century has sometimes gone hand in hand with the triumph of barbarism. Martin Luther King said it best: "Our scientific power has outrun our spiritual power. We have guided missiles and misguided men."

Among the humanists and the atheists it is somewhat of a truism to think of religion as a most destructive force in world history, and science, on the other hand, as a benevolent force. There is no doubt that in the name of religion endless horrors have been committed throughout history: wars between the Catholics and the Protestants, Holy Crusades and holy wars, devastating division of nations and countries on the basis of religion, the persecution of Jews, Sikhs, Hindus and of so many others. But all that pales before the nationalist, ideological and racial wars of the 20th century, often spurred by the genius and ingenuity of scientists and engineers.

Hitler, for instance, had no love for any religion but was intoxicated with the hatred of the Jews and overarching nationalistic ambitions. Stalin was an ideological tyrant and mass murderer, Japanese militarists were ruthless imperialists. Bangladeshis were murdered and raped not because they were not Muslims; Franco was a fascist who was ready to destroy half of Spain, if necessary, in order to wipe out the communists. Idi Amin was simply a tyrant who brooked no opposition. Dresden was fire-bombed not because its inhabitants were not Christians; the rape of Nanking happened not because of the religion of its people. The barbarism of 20th century is marked largely by ideology, by fascism and by abusive forces of nationalism.

In the history of Europe, we speak of the Dark Ages, of the Age of Enlightenment and Reason, of the Renaissance, of great sea voyages, of discovery of the New World, of imperialism and colonialism, of exploration and exploitation of the Dark Continent. All these and more have indeed been major steps in the evolution of European civilization, but time and again, Europeans, as indeed people in all parts of the word, have fallen far short of being 'civilized'. Mahatma Gandhi was once asked what he thought of European civilization. With his wry humour, he remarked: "It *would* be a good idea."

Civilization is indeed a very good idea. But how often, and how far, we are from it, is a numbing thought. If someone from outer space, or from another world, were to look at us humans objectively and give us a report card, surely they would conclude that we have often been – and continue to be – a very barbaric lot.

Take the twentieth century – the most barbaric century of all. It is still very much part of us, we are its children. In this century, in two world wars, in revolutions and civil wars, in coups and genocides, in wars for independence and rebellion, in atomic bombings and biological warfare, in forced famines and starvations, more than 187 million men, women and children were killed. That is almost six times the present population of Canada. By all accounts, on any scale of brutality and barbarism, that is a mind-boggling number. A million dead here, another million there, in Europe, in Asia, in Africa, everywhere. Schools and hospitals were blown up, bridges and museums were shattered, art and monuments were destroyed. Any semblance of civilization was a delusion. When one recalls the destruction of Tokyo, Hiroshima, Dresden, Nanking, Leningrad and Berlin, and the crematoria at Auschwitz, or the genocide of the Armenians in Turkey in 1915, or of the Bangladeshis in 1971, one can see how wild and preposterous these destructions had been.

The perpetrators of this barbarism sadly were great admirers of science and technology, and they supported their growth in every way. Bombings of Hiroshima and Nagasaki were political decisions, but the atom bombs were created by the genius and collaboration of thousands of brilliant scientists and engineers. Research on chemical and biological warfare was the work of scientists. The poison gases that came out of the showers at concentration camps in Germany and killed millions of Jews were not produced by ideologues but by scientists. Indeed scientists all over the world – including Robert J. Oppenheimer, the father of A-Bomb in America – were drawn to Stalinism and Communism, or to Fascism and Nazism, as a scientific experiment in social engineering. How can we forget that there were few amongst scientists – whether in the war machines of the USA, or of the Soviet Union or of Germany – who expressed any qualms about their efforts in the great nationalistic juggernaut of war. In the Military-Industrial Complex, which gave a new meaning to violence and wars, 'Science For Peace' became a faint cry.

In the disconcerting words of Australian physicist, Sir Mark Oliphant, who worked on the Manhattan Project for the production of the first A-bomb:

> I learned during the war that if you pay people well and the work's is exciting, they'll work on anything. There's no difficulty getting doctors to work on biological warfare, chemists to work on chemical warfare and physicists to work on nuclear warfare.

If all scientists, engineers and technologists were to take something like a Hippocratic Oath, of 'doing no harm' and 'treading with care in matters of life and death,' and abide by the oath as an ethical principle and moral law, an Age of Enlightenment may indeed arise on the earth.

❖

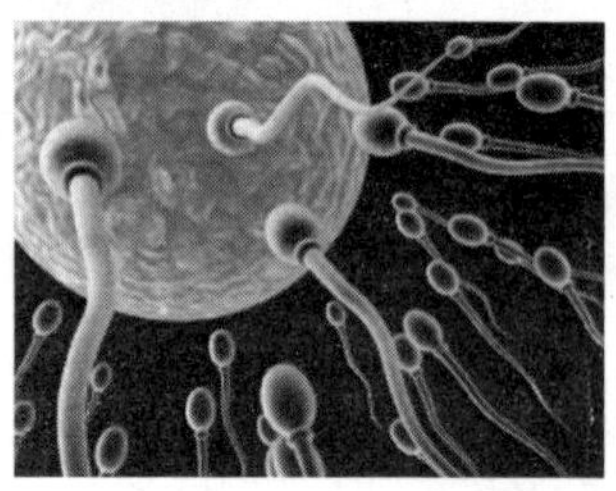

29

Peace or Violence in Our Genes?

In 411 BC, a remarkable play, *Lysistrata,* was performed in Greece. Written by one of the great playwrights, Aristophanes, the play presents a unique protest against the Peloponnesian War that was consuming men by the thousands. This war, like all wars since time immemorial, was enveloped in the frenzy of heroism and eternal glory. It was declared a necessary war; it was inevitable that blood be shed for honour and manhood.

The protest against the war in the play was in the form of a 'sex strike', in which the women withheld all physical intimacies from their husbands and lovers as long as they continued to indulge in the war. The protest proved to be successful; it seemed, at least in the play, that men had the sense, however briefly, to choose to make love rather than war.

I am not sure if such strategies have ever been tried in real life, but certainly there have been times when loud and eloquent voices have been raised against the absurdity and horrors of war. Songs have been composed, paintings created, plays and films made, novels written, all to expose the mindlessness of wars: how they are triggered, how they are manipulated and executed, how they are made to seem necessary and glorious, how they spread hatred, how the young are slaughtered like lambs in the war, and how wars never bring any resolution to any problem, and above all how they never bring peace. There are also others, though they have been humiliated and derided as cowards, who have refused to fight in wars on grounds of 'conscientious objection'.

In his book, *The Better Angels of Our Nature: Why Violence Has Declined*, psychologist Steven Pinker presents an interesting thesis: Despite the gruesome stories of violence everyday from all over the world, and despite the fact that in the twentieth century over 187 million died in wars, making it the most violent century in human history, over the last hundred thousand years, humans are becoming gradually but surely less violent and more peaceful.

Dr. Pinker presents elaborate data to prove how rule of law, a sense of justice and fairness, forces of modernity and other factors have contributed towards making revenge, an eye-for-an-eye, skirmishes and hostility, and wanton violence decline over time. In short, he argues, that peace is slowly breaking out. 'Give Peace a Chance', as John Lennon so famously pleaded, may possibly be getting a chance after all.

Is peace really becoming cool? Wars – hot and cold – have been cool and 'natural' for so long that it is somewhat difficult to believe that there may indeed be some bend in the human story.

As an ardent pacifist, this trend towards peace in human behaviour – if indeed it is true – is certainly most heartening to me. Yet there is an aspect of violence that is perhaps entirely new in the human history: transcending all ideologies and religious proclamations, waging wars is now an integral part of a military-industrial complex. Making more and more insidious weapons of mass destruction, and selling them to one and all, are new twists in the nature and industry of war. War is now a very, very big business. Was it so for Alexander, who went plundering the known world 2300 years ago? Was it so for the Romans or Genghis Khan? Was Napoleon driven by such commercial imperatives? Or Hitler or Stalin?

At what time in human history could a single weapon destroy a hundred thousand people, as happened in Hiroshima and Nagasaki? How could two young boys wipe away their fellow students and teachers, without any rhyme or reason, in a few minutes with a spray of bullets, as happened in Columbine High School in Colorado in 1999? How could a president of country fabricate stories to attack another country, as happened in Iraq, and send the whole world into an economic tail-spin?

The Wailing Wall' or 'The Western Wall' in Jerusalem is a sacred site for the Jews where, for centuries the faithful have prayed for peace. A story is told about a man who had prayed there for more than forty years, sometimes twice a day. His persistent devotion became matter of some interest to CNN and a reporter was sent to Jerusalem to cover the story:

"Is it true," the reporter asked the old man, "that you have prayed here for more than forty years?"

"Yes, it is true. Sometimes two or three times day."

"And what do you pray for?"

"I pray that Jews, Christians and Muslims would live in peace."

"And that goodwill will prevail amongst men and women of all faiths and ethnicities."

"And that the young will become the messengers of peace."

"So, how do you feel after all these years?" the reporter asked.

The old man replied: "I feel like I have been praying to a fu...ing wall."

With the futility of wars there has also been an equal sense of futility of prayers in stopping wars. There are many convoluted and nefarious reasons for it:

First, for millennia, wars have been promoted as necessary and integral to 'human nature'. Since wars and battles, or skirmishes and conflicts, have gone on in all human societies for so long, it seems to many – biologists, psychologists and others – that humans are 'wired' for violence, that we cannot do without it, and that it is in our genes. Inevitably, many of the biological studies – such as 'On Aggression' by Conrad Lorenz and 'Territorial Imperative' by Robert Ardrey – are based on study of animals, and then conveniently extrapolated to humans, making us completely rooted in the animal kingdom.

When it comes to humans, are we not both part of nature, and apart from it? What would civilization be otherwise?

"Nature," says Katherine Hepburn's Rose Sayer in the film *African Queen*, "is what we were put on earth to rise above."

Does our common animal ancestry condemn us to be no more than animals? Does our sharing of common genetic pool rob us of our humanity? What other animal uses 'drones' to kill and maim hundreds across continents while sitting thousands of miles away? What other animal has an arsenal of nuclear weapons to wipe away all of human civilization? Are we to be guided by our animal instincts, or by some higher values?

Second, wars are now a huge business; in the name of defence, 'preparedness for peace'; 'evil that exists in the world', nations, countries, tribes and communities have spent, and continue to spend, their most vital scientific, technological and human resources in creating more and more deadly weapons of mass destruction. For 2010, the ten largest defence budgets in the world totalled more than $1.1 trillion; the USA spent more than the next 15 countries put together. And all five permanent members of the Security Council of the UN – USA, Russia, France, Britain and China – have been in the business of selling weapons to everyone for years; they are indeed the great merchants of global death.

All countries – rich and poor, from Nepal to India, to Columbia, to China, to the USA – thus feed or are caught in the war frenzy, and create enemies, real and virtual.

There are many who accuse religions of contributing most to tensions and wars amongst people. But there is little mention of the scientists, engineers, mathematicians and biochemists and of their insidious contributions to the war effort. One may ask, are these millions of men and women, with exceptional skills and intellectual prowess, only passive, amoral pawns in the war machinery? Or are they its active attendants and instigators in the thievery of others' resources?

Third, for decades and centuries, everywhere in the world, wars and warriors have been glorified. Such was the emotional and patriotic glamour of wars that few people, if any, questioned the morality of wars, or explored any alternatives to resolving conflicts. Instead, poets wrote ballads in praise of mindless deaths of those who blindly followed military orders, however foolish, as Alfred Tennyson's famous poem, 'The Charge of the Light Brigade' about the suicidal attack in the Crimean War (1854-56) so glorifies:

Some one had blundered:
Theirs not to make reply,
Theirs not to reason why,
Theirs but to do and die.
....
When can their glory fade?
O the wild charge they made!
All the world wondered.

Over the decades, Tennyson's words have tended to take on a certain invocational élan, which, in popular parlance, came to be expressed as: "Ours is not to reason why, ours is but to do and die."

If one day, another group of sentient beings were to come to the earth, and examine human behaviour in wars, as we do now with bees or ants or wolves, they would probably find us a very uncivilized, self-destructive and barbaric lot. It is thus that I feel that anyone who attempts to pierce the impenetrable carapace of the glory of war and its false bravado is engaged in an heroic task. Those who have experienced the horrors of wars are the most ardent advocates for peace. In a song, "I Didn't Raise My Boy to be a Soldier," a mother expresses her pangs:

Ten million soldiers to the war have gone
who may never return again.
Ten million mothers' hearts
must break for the ones who died in vain.
...
I heard a mother murmur through her tears:
"I didn't raise my boy to be a soldier,
I brought him up to be my pride and joy.
Who dares to place a musket on his shoulder
to shoot some other mother's darling boy?"
Let nations arbitrate their future troubles.
It's time to lay the sword and gun away.
There'd be no war today if mothers all would say,
"I didn't raise my boy to be a soldier."

From Lysistrata in the fifth century BCE to 'The Mothers of the Plaza de Mayo' in Argentina in 1970s has been a long journey. But when women campaign to end Dirty Wars, the anguish of wars seethes in the marrow of our bones.

I wonder if the profiteers
Have satisfied their greed?
I wonder if the soldier's mother
Ever is in need?

I wonder if the kings
Who planned it all are satisfied?
They played their game of checkers
And eleven million died!
Oh, I'd like to see their faces
When they reach the Devil's door,
But even down in Hell
There is no torture such as war.

Eloquent and persistent voices are increasingly asking: Why do we fight? Why have we been fighting for so long and everywhere? The answers we are discovering are not hidden in some genes, or in some recesses of human nature, but in human folly and arrogance, and in politics and commerce. Whether through satire, as in the great novel, *Catch-22* by Joseph Heller, or in the film *Dr. Strangelove,* or through poignant portrayals of misery and horrors of war, as in the paintings of Francis Goya, and in such films as *Grave of the Fireflies, Ballads of a Soldier, No Man's Land* we begin to see, however hesitantly, that real heroism lies not in destruction but in creation, not in killing but in love, not in hatred but in compassion.

This is not a new discovery; the message of love has been proclaimed for as long as the trumpets of war have been blown. Perhaps we are no less wired for compassion than for indifference or callousness. It is obvious that in whatever manner they may be laid out, or woven, in the genes or in the human heart, the wires have been getting crossed again and again. Sometimes there is a spark, an illumination, or a stray light. But often, sadly, there is much darkness. Perhaps it is the role of art, of wise counsellors, and of courageous sojourners to remind (*re-mind*) us that each one of us is the weaver of our own destiny, that Nature is forever open to new possibilities; that there are always new tapestries ready to be woven and laid out.

❖

NOTES

p. 46 Charles Darwin, *The Descent of Man,* 1871, 1st edition, pp. 168-69.

p. 50 Dylan Thomas, poem: '*The force that through the green fuse drives the flower.*'

p. 94 Lars Malmstrom and David Kushner, Trs., *Four Screenplays of Ingmar Bergman*, Simon and Schuster, New York, 1960, pp. 111-12.

p. 95 Alok Bhalla, Tr., *Andha Yug: The Age of Darkness*, University of Hawai'i Press, 2010.

p. 106 Henry Wadsworth Longfellow, poem: *'A Psalm of Life.'*

p. 112 William Ernest Henley, poem: 'Invictus.'

p. 153 'Sir Mark Oliphant', Obituary, *The Economist*, July 20, 2000.

p. 158-59 'I didn't raise my boy to be a soldier' was an influential anti-war song in the pacifist movement in the U.S. before its entry in the First World War. The song inspired a sequel, several imitations as well as parodies.

p. 159 Billy Rose, "I wonder if the profiteers", *Best-loved Poems of the American People.*